Xiong Ming

Replace Lagrange multiplier method

AF154890

Xiong Ming

Replace Lagrange multiplier method

LAP LAMBERT Academic Publishing

Impressum / Imprint

Bibliografische Information der Deutschen Nationalbibliothek: Die Deutsche Nationalbibliothek verzeichnet diese Publikation in der Deutschen Nationalbibliografie; detaillierte bibliografische Daten sind im Internet über http://dnb.d-nb.de abrufbar.

Alle in diesem Buch genannten Marken und Produktnamen unterliegen warenzeichen-, marken- oder patentrechtlichem Schutz bzw. sind Warenzeichen oder eingetragene Warenzeichen der jeweiligen Inhaber. Die Wiedergabe von Marken, Produktnamen, Gebrauchsnamen, Handelsnamen, Warenbezeichnungen u.s.w. in diesem Werk berechtigt auch ohne besondere Kennzeichnung nicht zu der Annahme, dass solche Namen im Sinne der Warenzeichen- und Markenschutzgesetzgebung als frei zu betrachten wären und daher von jedermann benutzt werden dürften.

Bibliographic information published by the Deutsche Nationalbibliothek: The Deutsche Nationalbibliothek lists this publication in the Deutsche Nationalbibliografie; detailed bibliographic data are available in the Internet at http://dnb.d-nb.de.

Any brand names and product names mentioned in this book are subject to trademark, brand or patent protection and are trademarks or registered trademarks of their respective holders. The use of brand names, product names, common names, trade names, product descriptions etc. even without a particular marking in this work is in no way to be construed to mean that such names may be regarded as unrestricted in respect of trademark and brand protection legislation and could thus be used by anyone.

Coverbild / Cover image: www.ingimage.com

Verlag / Publisher:
LAP LAMBERT Academic Publishing
ist ein Imprint der / is a trademark of
OmniScriptum GmbH & Co. KG
Heinrich-Böcking-Str. 6-8, 66121 Saarbrücken, Deutschland / Germany
Email: info@lap-publishing.com

Herstellung: siehe letzte Seite /
Printed at: see last page
ISBN: 978-3-659-78331-9

Preface

This book is primarily written for students majoring in science and engineering, teachers in science and engineering, engineers, technicians, middle school teachers and philomaths. As you are reading this book, you will deeply appreciate the meaning of "mathematics is geometry", what geometric master Chen shengshen had said. This book clearly elaborates themes.

We correct a conclusion raised by Cantor Georg Ferdinand Ludwig Philipp more than 100 years ago in 1874, which has been existing present testbooks of mathematics all aroud the world. New approaches are presented in this book, which can replace Lagrange multiplier method raised by Joseph-Louis Lagrange more than 200 years in 1791, which has widely used in today's many fields of technique and engineering. The proof to non-uniue of Lagrange multiplier is given. Theorem of implicit functions is perfected and enriched the proposition of explicit functions by defined is proposted. New discrimination method for extreme value point under equality constraint condition also is presented. In this book, new meanings to calculating distances among point, curves and surfaces are given. But these questions can not be solved by Lagrange multiplier method. Mathematics teaching philosophy of *"Mathematics Around Us"* is proposed in this book.

If only audience have senior high school mathematics level, even if you don not know Cantor set theory or Lagrange multiplier mthod, you probably won't have any trouble reading capter one and chapter two of this book. But, if you have to be able to read the proof to the theorems in this book, you must have the basic knowledge of linear algebra and the basic knowledge of differential of multiple function, which are parts of mathematical course content of sophomore year of college.

The first draft of twenty pages in Chinese was finished in November 2013 year. In order to be published in journal, this manuscript was compressed as a paper in eight pages later. In order that on the basis of my research more scholars over worldwide can study further, I began to translate this work into English in mid-August this year. Athough it took me twenty days to translate this work into English, I think I have very good twenty days.

As an independent scholar, I carry out myself mathematical research work alone without any instruction, so my results are inevitably limited; I want other scholars to further study. I hope my mathematics research experience will offer the readers help .I also believe this book would be helpful to all readers.

If reader met any questions in your reading this book, you can contact me by email: Independent scholar: XIONG Ming Email: x599599@126.com.

Tel: 13882015920

2015.9.9   Chengdu China

Abstract

This book is primarily written for students majoring in science and engineering, teachers in science and engineering, engineers, technicians, middle school teachers and philomaths. The philosophy of *"Mathematics around us "* is proposed and the idea of correspondence is considered as a tool to deal with complex and profound uations. By interpretations for multivariate function, an error which exists in testbooks of mathematics for more than 100 years is corrected. After new approaches to solving extremum are developed. Lagrange multiplier method that is widely applied in mathematics, physics, engineering and technology will have fulfilled its historical task. Using n-dimensional outer product is, the dispute (if multiplier is unique or not) will end. Helping this new methods, we can calculate distance among point, curves, surfaces.

XIONG Ming graduated from Dept. Mathematics, Chongqing Normal University in 1991 year. He had engaged in junior (senior, college) mathematics teaching for more than 20 years. He published more than 20 papers about educational theory, mathematics education and mathematics research.

Keywords: correspondence; multivariate function; set theory; extremum; Lagrange multiplier; distance

TABLE OF CONTENTS

0.Introduction

In 2006, I started to teach higher mathematics and took up with element method and tried to solve multiple integral with element method. In the process of studing further multivariable calculus, I introduced an equation with a parameter in 2009 year, and proposed concepts of moving curve(surface), by which multiple integrals, except surface integral of the second type, can be reduced to simple integrals theoretically. Readers may refer to XIONG Ming: *Construction of the Regional Elements of Differential Calculus and Re-integration into Sigle-integral Directly*.doi:10.3969/j.issn.1000-5757.2009.08.032. *Change Double Integral into Simple Integral with Area Element*, and *Moving Curve (Surface) and Degenerate Transformation for Multi-integral*.

What is more, I ascertained many new interpretations for multivariate function, which can not only be used to interpret the processe of converting multiple integrals into simple integral directly and the processe of solving conditional extremum problem, which can but also be used to discover the conclusion of Cantor set theory to be wrong.

The Mathematics teaching philosophy of *"Mathematics Around Us"* is proposed in this book. By introducing a monotonic and continuous function, a one-one correspondence between two intervals is build; and by using parametric equations of the curve, a one-one correspondence from the points on the curve to the points on the line is established. Specially, the meanings of multivariate functions are given. By using a n-variable equation with a parameter, a correspondence from n-dimensional space area to a interval is built, so the wrong conclusion is completely denied. This enriches calculus and can reduce the teaching difficulty of real function in some degree. The expression of moving curve (surface) limit is given in thisbook.

Starting from the geometric meaning of the extreme point of unary function and the geometric meaning of the ordinary extreme point of binary function, and being combined with the knowledge of space vector, new methods gradually are derived. Namely, using $(n \times n)$ system of equations, the extreme points of function of n-variables under m-equality constraints can be acquired. We rebuild the theories and approaches which can easily solve extreme problem of multivariate function under equality constraints.The new approaches can eliminate the disadvantages and shortcomings of Lagrange Multiplier method .In the process of re-developing new theories, by using outer product in n-dimensional space, we define general equation of curve in n-dimensional space and its directional vector and we present the geometrical significance for the extreme points of function of n-variables under the condition of one-equality constraint or under the condition of n-1 equality constraints .From the simple to the complex ,from the particular to the general to the more general, we develop the new approaches, which have more advantages over Lagrange Multiplier method and can absolutely replace it. Using outer product in n-dimensional space, we also demonstrate the expressions of Lagrange Multipliers, which ends the dispute (if Lagrange Multiplier is unique or not) between Qian Weichang and Hu Haicang for 30 years. The derivation process of the new approaches involve moving curve (surface), knowledge of vector, implicit function theorem, outer product and matrix operation, etc. This work colligates geometry, algebraic and analysis. The new approaches can easily calculate distance among point, curves and surfaces.

By the way, I also proposed that outer product dose not exist only in three dimensional space or seven dimensional space, it actually exists in any dimensional space, and it has extensive applications. Reader may refer to LIU Hai-feng: *On the Extension of Vector Product in Multi-dimwnsional Euclidean Space* in *College Mathematics*1672-1454(2014)03-0074-05 and XIONG Ming: *Generaliztion of Outer product and Application* in *Sdudies in college Mathematics* 10.3969/j.issn.1008-1399.2014.04.018)

1. "Mathematics Around Us"

According to materialist dialectics ,all things in the real world are closely releted to each other .Mathematics is no exception ,any mathematics concept comes from real life ,mathematics and our life have innumerable ties , Matheamtics widely exist in our real life ,any mathematics concept is associated with our real life ,namely "*Mathematics Around Us*" .The philosophy of "*Mathematics Around Us*" helps to cultivate studens' abilities of observing and analyzing and help gradually develop creative thinking of students ,If mathematics educators adhere to this mathematics teaching philosophy ,mathematics teaching difficulty is not only reduced greatly ,student's mathematics interest is but also stimulated .

Modern science and technology all are based on mathematical theory ,function theory is the most basic theory of many mathematical branches ,and correspondence is one of the most foundational concepts of mathematics .Nevertheless, the status and importance of correspondence in mathematics is offen ignored by us . As a matter of fact ,correspondence is an instrument to deal with many complex and profound quesiton ,it can make some complex maters simplify and help us visualize the process of comlex though ;correspondence is also an important idea in mathematics .But there are few detailed explanation for the concept of correspondence in current mathematics textbooks ,which has a direct effact student's understanding of functions ,and make some students generally arise the feeling of being feared by mathematics ..

Using the idea of correspondence ,Leonhard Euler(1707-1783) made each piece of land correspond the only point ,each bridge correspond the only line ,he dealt with Seven Bridges Problem very tactfully ,which resulted in graph theory. Thereby ,by utilizing the idea of correspondence rationally ,one can more easily solve some profound and complex problems .

Correspondence ,in short ,a relationship between elements in a set and elements in other set ,corresponding law is the rule or regulation of this relationship .

"*Mathematics Around Us*" ,we firstly give a familiar example .

Suppose the price of a certain apple sold in a supermarkets is 10 RMB per kilogram ,then "10 RMB per kilogram" is a corresponding law ,the weight of apple that you buy is corresponding the param money .

Likewise, "take the square root of positive real number" is also a correspondence law ,the corresponding relation is expressed as $y = \pm\sqrt{x}, x \in R^+$. Correspondences are divided into three categories :one-one correspondence ,many-one correspondence and one-many correspondence .   We will only discuss the first two correspondences here

1.1.The definition of one-one correspondence

The origin,per unit length and positive direction are determined on a line,every real number corresponds to the only point on number axis ,every point on number axis corresponds to the only real number.Under rectangular coordinate system ,each point in the plane corresponds to only the ordered pair(a,b) ,and each ordered pair (a,b) corresponds the only point in the plane .The class of correspondences are called one-one correspondences.

mapping: Suppose A and B are two sets. According to a certain corresponding rule f ,if each element a in A has the only element b in B ,the rule f is said to be mapping from A to B , which is expressed as $f : A \rightarrow B$, where a in A is called a preimage, and b in B is called an image.

one-one correspondence: Suppose $f : A \rightarrow B$ is a mapping from A to B .If each element b in B

has the only preimage a,the mapping $f:A \to B$ is called one-one correspondence(or one-one mapping,one-one transformation).

Any function is a mapping from domain to value region.An independent is a preimage,a function value is an image .Monotone functions are one-one correspondences.Quadratic functions are two-one mappings.Periodic functions are "infinitely many-one" mappings.

1.2.One-one correspondence from an interval to the other interval

Using $y = \dfrac{12}{5}x$,Bolzano (Czekh,1781-1849) constructed one-one correspondence from the interval[0,5] to the interval [0,12] .He proposed that there were the same many in set A as that in set B ,if a one-one correspondence existed in between A and B.

Actually,when the two intervals are regard as the domain and the value region separately, using a appropriate monotone and continuous function,the one-one relationship is established

Example 1 Establish one-one correspondences from an interval to the other interval.

(1) $[a,b) \to [c,d)$,

(2) $(-\infty,+\infty) \to (0,1)$,

(3) $(0,1] \to [0,+\infty)$,

(4) $[a,b] \to (-\infty,+\infty)$, $[a,b]$ is any interval.

Solution: (1)$[a,b)$ and $[c,d)$ are regard as the domain and the value region separately,then the one-one correspondence is established by

$$y = \frac{d-b}{c-a}(x-a)+b .$$

(2) $y = \pi^{-1}arc\cot x, x \in R, y \in (0,1)$.

(3) $y = \log_{0.5} x, x \in (0,1], y \in [0,+\infty)$.

(4) Firstly,we determine the altitude points A and B right above a,b on number axis and drow a smooth and concave curve L between A and B .Secondly, we draw a ray line from the midpoint P of AB to intersect L and the number axis at M and N separately.Thus M corresponds to N ,and the horizontal coordinate of M corresponds to N on number axis.So,the correspondence is eatablished between $[a,b]$ and $(-\infty,+\infty)$.

Strictly speaking,there is no one-one correspondence between open-interval and closed interval unless the two endpoints on closed interval are removed,and it is as well between open interval and half-open interval or between closed interval and half-open interval.

As these examples illustrate :there are as many real numbers on R as on any open interval of R ,as well as between any interval and the line.

1.3.One-one correspondence between a curve and the line.

Every point $(x, f(x))$ on the graph of a function $y = f(x)$ correponds the only point x on the domain.If the function is continuous,the graph of it is a continuous curve .For any given curve,by using geometrical method or algebraic means,the one-one relation is established between the points on the curve and on the line,which can show that there are the same points on the curve as on the line.

Example 2 The points on a circle are the same many as the points on the line.

Proof :Method-1 .Let a circle touch the number axis at the origin(the tangent point corresponds to the origin).Let a tangent line from any point A on the circle intersect the number axis at a.(Set the tangent point corresponds to $-\infty$ or $+\infty$,when the tangent line and the number axis are parallel to each other),the points on the line are the same many as on the line.Because any line segment can be formed a circle ,the points on a circle are the same many as on the number axis ,the pints on a circle are the same many as on the line .

In fact ,by using a method like Method-1 ,we can prove that the points in a sphere are the same many as the points in the plane .

Method-2 The function $u = 2\pi - 2arc \cot t$ establishes the one-one correspondence from $(-\infty,+\infty)$ to $(0,2\pi)$,and the parameter equations

$$\begin{cases} x = \dfrac{R}{\sqrt{6}}\sin u + \dfrac{R}{\sqrt{2}}\cos u \\ y = \dfrac{R}{\sqrt{6}}\sin u - \dfrac{R}{\sqrt{2}}\cos u \\ z = -\dfrac{2R}{\sqrt{6}}\sin u \end{cases}$$

establishes the one-one correspondence from interval$(0,2\pi)$ to the circle

$$\begin{cases} x^2 + y^2 + z^2 = R^2 \\ x + y + z = 0 \end{cases} ;$$

So the following system of euations with a parameter

$$\begin{cases} x = -\dfrac{R}{\sqrt{6}}\sin(2arc \cot t) + \dfrac{R}{\sqrt{2}}\cos(2arc \cot t) \\ y = -\dfrac{R}{\sqrt{6}}\sin(2arc \cot t) - \dfrac{R}{\sqrt{2}}\cos(2arc \cot t) \\ z = \dfrac{2R}{\sqrt{6}}\sin(2arc \cot t) \end{cases}$$

establishes the one-one correspondence from $(-\infty,+\infty)$ to a circle.

Note:The point $\left(\dfrac{R}{\sqrt{2}}, -\dfrac{R}{\sqrt{2}}, 0 \right)$ correspondes to $-\infty$ or $+\infty$.

Because the points on any curve are the same many as the points on the line ,and the line is one-dimensional spae ,any curve is one-dimensional space .

1.4.One-one correspondence between rational numbers on any interval and positive integers in the set of positive integers .

Cantor proposed that a set was called countable if the all elements in the set could be arranged at certain regular.He divided all the sets into two types:countable sets and uncountable sets,and proposed that the set of integers and the set of rational numbers were all countable,and there were as many positive rational numbers   as positive integers .The set of irrational numer and the set of real numers are all uncountable .In spite of   infinitely many rational numbers and real numbers between any two   real numner, each rational number can be expressed as the quotient of two integers ,thus rational numbers can be arranged at a certain manner. But,any irrational number is a non-repeating and non-terminating decimal number,and anyhow cannot be expressed as the quotient of two intrgers,it can only be the arithmetic of adding,subtracting,multiplying,and dividing for infinitely many integer numbers ,or it can only be the limit of a sequence of rational numbers. Irrational number on $(-\infty,+\infty)$ can not be arranged through any means .The above all are the differences and relations.

Example 3 There are the same rational numbers on any interval as positive integers.

Proof :For convenience,we will first prove that the number of rational numbers on interval (0.1,0.2) is equal to number of positive integers.

We arrange the all rational numbers on the interval under the following rules:

(1) ascending denominator,

(2) ascending numerator if the denominator is the same ,

(3) keepping the former if a rational number is equal to the former.

So ,the all rational numbers on the interval (0,1,0,2) is arranged as follows :

$$\frac{2}{11}, \frac{2}{12}, \frac{2}{13}, \frac{2}{14}, \frac{2}{15}, \frac{2}{16}, \frac{3}{16}, \frac{2}{17}, \frac{3}{17}, \frac{2}{18}, \frac{2}{19}, \frac{3}{19}, \frac{3}{20}, \frac{3}{21}, \frac{4}{21}, \frac{3}{22}, \frac{3}{23}, \frac{4}{23}, \cdots\cdots$$

As the denominators are increasing and tending to infinity ,the all rational numbers on(0.1,0.2) are lined up as a sequence .Each rational number on the interval corresponds to the only one positive integer. The all rational nubers on any interval are similarly arranged in the above maner.

The number of elements in a set is called the cardinal number or cardinality of the set.Any interval is a continuum ,C is used to express the cardinality of any interval.

2.Interpretations for multivariate functions

2.1.About equation $F(x_1, x_2, ... x_n) = t$

The system of coordinates has built a bridge between algebra and geometry .Any binary equation with infinitely many solutions represents a curve(line ,half-line,polyline) ,for instance y=|x| denotes a V-type curve .Any continuous function of one variable is binary equation and denotes a curve .Similarly,any continuous function of two variables is a ternary equation and denotes a surface ,for instance $z = x^2 + y^2$ denotes a paraboloid of revolution.In the process of changing multiple integrals or surface integrals directly into simple integrals ,an eqaution with a parameter .

$$F(x_1, x_2, ... x_n) = t \qquad ①$$

is introduced,which is called to be a degenerate transformation. $F(x_1, x_2, ... x_n) = t$ expresses a moving surface(when n=2,it expresses a moving curve),and is used to describe or simulate continuous variation of object.But ,It has many significances.

2.1.1 The meanings of equation $F(x, y) = t$

Example 4 Give the meanings of the following equations

(1) $y = tx^2$ ②

(2) $xy = t(t > 0)$ ③

Solution:(1)When t is assigned a non-zero real number,a parabola is obtained from the equation②.When $t = 0$, $y = 0$,② expresses x-axis.When t has travered all the real numbers,a family of parabola in the plane is obtained,which have a common apex and a common axis of symmetry.Each parabola and all points on it correspond to the only real number,and vice versa.So,equation② establishes a correspondence from the coordinate plane to the interval $(-\infty, +\infty)$.When t is continuously changing on $(-\infty, +\infty)$,② represents a moving parabola.The smaller $|t|$ is, the large the opening of the moving parabola is.When $t \to 0$,the moving parabola is infinitely tending to x-axis;when $t \to \infty$,the moving parabola is tending to y-axis. Using symbols of set and limit,the two ultimate states are expressed as respectively:

$$\lim_{t \to \infty}\{l : y = tx^2, t \in R\} = \{l_1 : x = 0\},$$

$$\lim_{t \to 0^+}\{l : y = tx^2, t \in R\} = \{l_2 : y = 0\}.$$

(2) The equation③ $xy = t(t > 0)$ denotes a family of hyperbola in the first and the third quadrants.They have common symmetry axis(line $y = x$) and a common asymptote(the x and the y coordinate axes),the vertex of any hyperbola is on the line $y = x$ .Each hyperbola and the points on it are correspond to the only real number,and vice versa. ③ establishes a mapping from the first quadrant and the third quadrant to the interval $(0, +\infty)$.

When $t \to +\infty$, the moving hyperbola is moving to infinity along the line $y = x$ .But,when $t \to 0^+$, the moving hyperbola is tending to the two axis, which is the limit state of the moving curve and is expressed as

$$\lim_{t \to 0^+}\{l : xy = t, t > 0\} = \{l_0 : \text{ the two axis }\}.$$

So,the equation $F(x, y) = t$ expresses a moving curve and a family of curves in the plane,and a mapping

2.1.2. Meanings of equation $F(x, y, z) = t$

Analogously,any equation $F(x, y, z) = t$ has three meanings:a family of surfaces ,a moving surface and a mapping.

Example 5 Give the meanings of the equations

(1) $x + y + z = t(t \in R)$ ④

(2) $x^2 + y^2 + z^2 = t(t > 0)$ ⑤

(3) $xyz = t$ ⑥

(4) $\dfrac{x^2}{a^2}+\dfrac{y^2}{b^2}-\dfrac{z^2}{c^2}=t$ ⑦

Solution:(1)④acts as a moving plane and a family of planes in the space; a mapping that makes each plane and the infinite points in the each plane　correspond to the only real number,and vice versa.

(2)⑤ acts as a moving sphere and a family of spheres in the space;a mapping that makes each sphere and the infinite points in each sphere correspond to the only positive number,and vice versa.

(3)When t is assigned a non-zero real number,a four-leaf surface is obtained from the equation;when t is assigned a positive real number,the four-leaf surface is located 1,3,6,8 four octants;when t is assigned a minus,the four-leaf surface is located 2,4,6,8 four octants.

For convenience,the case in 1 octant is discussed.When t has gone through real numbers on $(0,+\infty)$, $xyz=t(x,y,z,t>0)$ expresses a family of surfaces,which have a common asymptotic surface (three coordinate planes in 1 octant),all the vertex $(\sqrt[3]{t},\sqrt[3]{t},\sqrt[3]{t})$ on the line $x=y=z$.

The equation $xyz=t(x,y,z,t>0)$ establishes a mapping from 1 octant to the $(0,+\infty)$, which make each surface and the infinite points in the each surface correspond to the only positive number.

When $t\to+\infty$,the moving surface is moving　tending towards infinity along the the line $x=y=z$ .When $t\to 0^+$, it is tending to the three coordinate planes in the 1octant ,which is the limit state of the moving surface.It is expressed as

$$\lim_{t\to 0^+}\{\Sigma:xyz=t,x,y,t>0\}=\{\Sigma_0:\ \text{three coordinate planes in the 1octant}\}.$$

(4)⑦denotes a moving hyperboloid and a family of hyperboloids in the space ;a mapping that makes each hyperboloid and the infinite points in each hyperbolobid correspond to the only number ,and vice versa .When $t\to 0$, it is tending to the cone $\dfrac{x^2}{a^2}+\dfrac{y^2}{b^2}=\dfrac{z^2}{c^2}$,which is the limit state of the moving surface .It is expressed as

$$\lim_{t\to 0^+}\{\Sigma:\dfrac{x^2}{a^2}+\dfrac{y^2}{b^2}-\dfrac{z^2}{c^2}=t,t\in R\}=\{\Sigma_0:\dfrac{x^2}{a^2}+\dfrac{y^2}{b^2}=\dfrac{z^2}{c^2}\}.$$

2.2.The meanings of multivariate functions

An equation $F(x,y)=t$ is actually a function of two variables $z=F(x,y)$. Thus,the meanings of a function of two variables $z=F(x,y)$ are as follow :a surface in space, a moving curve ;　a family of curves;a mapping from the domain to the value region of the function ,which makes each curve and all the points on the each curve correspond to the only value of the function.

Functions of a variable can be classified into two kinds:one-one correspondence ,several-one correspondence .But, any function of two variables is a infinity-one correspondence(the extreme points and the points of maxima or minima can be excepted).

Choose any function value z_0 (non-extremum and non-maximum ,non-minimum),then $F(x,y)=z_0$ is a planar curve ,on which the infinite points are corresponding to the only z_0 .

Owing to infinitely many real numbers between any two real numbers,there are infinitely many curves between any two curves

$$F(x, y) = z_1, F(x, y) = z_2 ,$$

so the domain of the function $z = F(x, y)$ is completely covered with a family of curves $F(x, y) = t$.

Similarly, $F(x, y, z) = t$ is a function of three variables $u = F(x, y, z)$. Thus,the meanings of a function of three variables $u = F(x, y, z)$ are as follows :a moving surface and a family of surfaces in the space;a mapping from the domain to the value region of the function, which makes each surface and all the points in the each surface correspond to the only value of the function.The domain of the function $u = F(x, y, z)$ is completely filled with a family of surfaces.

But, any function of three variables is a infinite many-one correspondence(the extreme points,the points of maxima or minima are excepted).

A function value u_0 (non-extremum and non-maximum ,non-minimum)is given,then $F(x, y, z) = u_0$ is a surface in which the infinite points correspond to the only u_0 .

Generally, a continuous function of two variables expresses a surface in the space ;a degenerate transformation;a moving curve in the area; a family of curves that cove completely the whole area;a mapping from the area to the value region of the function ,which makes each curve and all the points on the each curve correspond to the only value of the function. Any continuous function of n-variables(n$\geqslant$3) expresses a moving surface in n-dimensional space;a degenerate transformation;a family of surfaces that fill completely in the n-dimensional domain;a mapping from the n-dimensional domain to the value range of the function,which makes each surface and all the points on the each surface correspond to the only value of the function.Any multivariate function is a infinite many-one correspondence from the points of the domain to the points of the functional value.

Furthermore, because any binary equation with infinite many solutions expresses a curve ,and a curve is a one-dimensional space,so the binary equation expresses a one-dimensional space,and a function $z = F(x, y)$ expresses a moving one-dimensional space and a family of one-dimensional spaces in the domain of the function. Similarly,any ternary equation with infinite many solutions expresses a surface,and a surface is a two-dimensional space,so the ternary equation expresses a two dimensional space,and a function $u = F(x, y, z)$ expresses a moving two dimensional subspace and a family of two dimensional subspaces in the domain of the function .By extension,any function

$$z = F(x_1, x_2, ..., x_n)(n \geq 2)$$

expresses a moving n-1 dimensional subspace and a family of n-1 dimensional subspaces in the domain of the function, and the family of n-1 dimensional subspaces correspond to the points on the rang of the function of n variables.The cardinality of a family n-1 dimensional subspaces in n-dimensional space is equals to the cardinality of any interval.

2.3.Correction for Cantor' Conclusion and Significances

2.3.1.A wide spread conclusion of Cantor set theory

In 1874 ,Cantor started to consider whether there is a one-one correspondence between the points on the line and the points in the n-dimensional space .After 3 years' hard work ,he said proudly that "*there is a one-one correspondence between n-dimensional space and a line ,and the points on n-dimensional space are as many as on a line*" in his private communication with Dedekind .Since Cantor set theory had been published ,it has been attracting extensive attention from mathematicians ,philosophers ,logicians, Cantor received strong support from authorities such as Dedekind,Weierstrass, Mittag-Leffler .Hilbert had praised highly of Cantor and rated continuum hypothesis the top of the the famous 23 probems at the 2nd ICM in 1900 ,mathematicians all over the world have taken the conclusion for granted .To this day ,hardly anyone questions the conclusion ,it has appeared in numerous textbooks of high school maths, mathematical history,and the equivalent proposition of the conclusion:

Suppose the sign C denotes the cardinality of continuum(an interval or the line) ,then the cardinality of E^n

or E^∞ (n-dimensional space or infinite dimensional space) is still C .

has appeared in numerous textbooks of set theory and real variable function,etc.

2.3.2.Correspondence between n-dimensional space and line

From the perspective of correspondence and geometry, any multivariate function is one-one correspondence and is also infinitely many-one correspondence .

For any n-dimensional domain and any interval ,let the domain be the domain of a function of n variables and the interval be the rang of the function ,a one-one correspondence from a family of n-1 dimensional subspace to the interval is established by the function .As long as a one-one correspondence is established between set A and set B ,the elements in A are the same many as that in B .Due to the one-one correspondence from any interval to the line ,and owing to the one-one correspondence from a family of n-1 dimensional spaces in n-dimensional space to an interval, there is a one-one correspondence from a family of n-1 dimensional subspaces to the line ,but there are infinite points on each n-1 dimensional subspace ,thus the conclusion that there is a one-one correspondence between n-dimensional space and the line ,and the points on n-dimensional space are as many that as on the line is obviously mistaken.

2.3.3.Correction for the conclusion of Cantor set theory

We correct Cantor' conclusion as the followings:

There is a one-one correspondence between any interval and the line ,the points on any interval are as many as that on the line .There is a one-one correspondence between any curve and the line ,the points on any curve are as many as on the line.There is a one-one correspondence from a family of curves that cover completely the plane to the line ,the curves are as many as pints on the line.There is a one-one correspondence from a family of surfaces which fill completely the 3-dimensional space to the line,the surfaces in 3-dimensional space are as many as the points on the line.There is a one-one correspondence from a family of n-1 dimensional subspaces which fill in the n-dimensional space to the points on the line.There is no one-one correspondence from the points in n-dimensional space to the points on the line.

Let $\overline{\overline{}} \cdot$ express the cardinality of a set ,then

$$C = \overline{\overline{E}} < \overline{\overline{E^2}} < \overline{\overline{E^3}} < ... < \overline{\overline{E^n}} < ....$$

Cantor developed point-set theory ,and exploited a new world—set theory .Based on Cantor set theory ,many scholars have made great achievements ,Although Cantor set theory is not perfect ,he is still one of the greatest mathematicians in the history of mathematics .

2.3.4.Significances of Correcting Cantor' Conclusion

(1) Correction for Cantor' conclusion can reduce teaching difficulty largely

By excavating the teaching convergence between middle school mathematics and college mathematics ,the teaching difficulty is largely reduced (reader may refer to *Soving Equation of Common Vertical Line of Two Lines in different Planes with High School Mathematics*),which is what mathematics educators ought to do .The all above can not only strengthen the relations between middle school mathematics ,calculus ,space analytic geometry ,real function theory ,can but also reduce the teaching difficulty of real analysis in some degree .

(2)Correction for Cantor' conclusion can result in new mathematical approaches

After proposing the concepts of moving curve(surface) and the concept of family curves(surfaces) ,and revealing the relations between moving curve(surface) and multivariate function ,we can give the geometric meanings and the physics meanings of the problem of conditional extreme value .New approaches which are superior to current Lagrange multiplier method are obtained can completely replace it .Based on the concepts of moving curve(surface) and set theory ,the method of changing directly multiple integrals or first type surface integrals into simple integrals can have visual geometric meanings and physics meanings ,and it is absolutely necessary to redefine multiple function integrals.

3. Imperfection of Lagrange Multiplier Method

Optimization probems in real life (such as maximization of profit and minimization of cost , etc) or in science and technology often can be converted into extremum problems of functions .In mathematics optimization， the method of Lagrange multipliers is considered to be a strategy for finding the local maxima and minima of a function subject to equality constraints.The method is used in economics ,optimal control theory ,nonlinear programming ,enven in military theories ,etc. In seeking the solution to the fastest falling curve ,by helping pure analytical oproach ,Leonhard Euler(1707-1783) and others developed variational method .Joseph Louis Lagrange(1735-1813) was one of the creators of calculus of variations .Deriving the Euler-Lagrage equations for extrema of functionals ,Lagrage extended the method to take into account possible constraints ,and arrived at the method of Lagrange Multipliers in 1791 .Qian Weichang(1912-2010) ,Hu Haicang(1928-2011) and others developed generalized variational method ,which is used in cybemetics in the 1980s ,.

But, Lagrange Multiplier Method has inherent imperfections . *"Mathematics Arround Us"* ,in order that everybody can understand the deficiencies of Lagrange Multiplier Method ,we first give real life examples to let us see how the method solves conditional extreme values of a function subject to equality constraints.

Question one:Suppose there are 10-meter slates.The slates are used to construct a flowerbed near a wall .Make the area of the flowerbed as large as possible. What is the length?What is the width?

Analysis: Suppose the length and width of the flowerbed is x, y separately ,then the area

$$S = xy \tag{1}$$

While we solve the largest area of the flowerbed, x, y must satisfy following conditions

$$\begin{cases} x > 0, y > 0 \\ x + 2y = 10 \end{cases} \tag{2}$$

This is conditional extremum problem of binary function.

Of course ,the conditional extremum problem can be changed into the following extremum problem of the quadratic function

$$S = \frac{1}{2}x(10 - x), 0 < x < 10$$

But many extremum can not be changed into extremum problem of univariate function .

Question two:Please design an open water tank in retangle shape ,which volume is the definite value V . Make the surface area of the water tank as small as possible. What are the dimensions of the water tank?

Analysis: Suppose the length, width and height of the water tank is x, y, z separately ,then the surface area of the water tank

$$S = xy + 2(xz + yz) \tag{3}$$

While we solve the smallest area of the water tank, x, y, z must satisfy following conditions

$$\begin{cases} x, y, z > 0 \\ xyz = V \end{cases} \tag{4}$$

Extremum problems such as question one and question two are called extreme value problem with constraint condition. Functions (1) and function (3) are called objective functions，condition(2) and condition (4) are call to be the corresponding constraint conditions.

By using Lagrange Multiplier Method to solve the above two problems ,if one must introduce a new variable and establish a new function of three variables and a new function of four variables respectively

$$u(x, y, \lambda) = xy + \lambda(x + 2y - 10),$$

$$u(x, y, z, \lambda) = xy + 2(xz + yz) + \lambda(xyz - V),$$

where λ is an undetermined constant ,which is called to be Lagrange Multiplier . In order to obtain extreme points, if according to Lagrange Multiplier method ,we must take the partial derivatives of function of three variables and function of four variables respectively ,then let the all partial derivatives be zero ,and then solve system of termary equations and system of quadruple equatios respectively.

More generally , according to Lagrange Multiplier method, if one solves conditional extreme value of function of n variables under $m(m < n)$ equality constraints ,one has to take $n + m$ partial derivatives and

has to solve $(n + m) \times (n + m)$ system of equations.

As we can see from Lagrange Multiplier method ,after introducing m new variables to establish new function of $n + m$ variables, it will be more difficult to obtain the points of extreme value ,especially when $n, m > 3$ .This is the imperfection of Lagrange Multiplier method .Fortunately,new approaches are obtained by us.

4. Basic knowledge of Derivation of the New Approaches

Before the new approaches are presented ,we must familiarize the following basic knowleges.

4.1.The meanings of Multiple Equation

An equation $F(x,y) = C$ which has infinitely many real solutions expresses a planar curve ,and a cylindrical surface which is perpendicular to coordinate surface xoy .An eqution $F(x,y,z) = C$ which has infinitely many real solution expresses a surface in space.

4.2.A family of Curves in the Plane and a Moving Curve in the Plane

The equation with a parameter $F(x,y) = t$ is called a family of curves in the plane and or a moving curve in the plane .The equation with a parameter $F(x,y) = t$ is different from common equation $F(x,y,z) = 0$,although the equation $F(x,y) = t$ has also three variables ,where the variable t varies independently from the two variables x,y ,variation of t causes chang of x,y .

4.3.A Family of Surfaces and a Moving Surface

The equation with a parameter $f(x,y,z) = t$ is called a family of surfaces or a moving surface .The equation with a parameter $F(x,y,z) = t$ is different from common equation $F(x,y,z,w) = 0$,although the equation $F(x,y,z) = t$ has also four variables ,where the variable t varies independently from the two variables x,y,z .

4.4.Partial Derivative

A partial derivative of a function of several variables is its derivative with respect to one of those variables ,with the others held constant.

The partial derivative of a function $f(x,y,z)$ with respect to the variable x is variously denoted by

f'_x, f_x or $\dfrac{\partial f}{\partial x}$.

In the process of computing their partial derivatives ,the derivative rule and derivative formula for function of one variable are still used .For example, the partial derivative of the function $z = x^2 y + xy^3$ with respect to x and y are $z_x = 2xy + y^3, z_y = x^2 + 3xy^2$.

4.5.Tangent Vetor and Nomal Vector to a Curve

If the slope of a straight line is k ,then the direction vector of it is the vecor $(1,k)$ and the normal vector of it is the vector $(k,-1)$ .A continuous function of one variable $y = f(x)$ denotes a curve, if the function is derivable at the point x_0 , then the direction vector and normal vector of the curve at x_0 are $(1, f'(x_0)),(-f'(x_0),1)$ respectively .The vector $(1, f'(x_0))$ is perpendicular to the vector $(-f'(x_0),1)$.

In fact, a function $y = f(x)$ is a binary equation $f(x) - y = 0$ and a binary equation denotes a curve .So the expression of the normal vector to a planar smooth curve $F(x, y) = C$ is $N = \pm(F_x, F_y)$, and the expression of the direction vector is $v = \pm(F_y, -F_x)$ .For example ,the expresse of normal vector to the smooth curve $x^2 + y^4 = 4$ is $N = \pm(2x, 4y^3)$ or $N = \pm(x, 2y^3)$,and the exprsse of dirction vector is $v = \pm(2y^3, -x)$.

4.6.Normal Vector to Surface

The expression of the normal vector of a smooth surface $F(x, y, z) = C$ is $N = \pm(F_x, F_y, F_z)$.

For a plane given by the equation

$$ax + by + cz = d ,$$

the normal vector is

$$N = \pm(a, b, c) ;$$

and for a surface given by the equation

$$\frac{x^2}{a^2} + \frac{y^2}{b^2} + \frac{z^2}{c^2} = 1 ,$$

the normal vecor is $N = \pm(\frac{x}{a^2}, \frac{y}{b^2}, \frac{z}{c^2})$.

4.7.Collineation ,Coplane , Linear Dependence of Vector

Suppose $\vec{a}, \vec{b}$ are two vector ,where $\vec{b} \neq \vec{0}$ .If there exists a real number λ which satisfies $\vec{a} = \lambda \vec{b}$,then $\vec{a}, \vec{b}$ are called collinear or parallel.

Suppose $\vec{a} = (a_1, a_2, ..., a_n), \vec{b} = (b_1, b_2, ..., b_n)$ .If $\vec{a}, \vec{b}$ are collinear , then

$$\frac{a_1}{b_1} = \frac{a_2}{b_2} = ... = \frac{a_n}{b_n} .$$

Suppose $\vec{a} = (a_1, a_2, a_3), \vec{b} = (b_1, b_2, b_3), \vec{c} = (c_1, c_2, c_3)$ are three vector.If determinant

$$\begin{vmatrix} a_1 & a_2 & a_3 \\ b_1 & b_2 & b_3 \\ c_1 & c_2 & c_3 \end{vmatrix} = 0 ,$$

then three vector are called coplanar, three vector are also called linear dependent.

When the three vectors are coplanar ,a vector of the three vectors can be expressed as linear combination of the remaining vectors .In other words ,if there exist three real number $\lambda_1, \lambda_2, \lambda_3$ which satisfy

$$\lambda_1 \vec{a} + \lambda_2 \vec{b} + \lambda_3 \vec{c} = \vec{0}$$,then the three vectors are linearly depenternt .

In general ,a set of vectors is called to be linearly dependent if one of the vectors in the set can be defined as a linear combination of the other vectors.

Namely ,suppos $\alpha_1, \alpha_2 ..., \alpha_n$ are set of vectors .If there exist n real numbers $\lambda_1, \lambda_2, ..., \lambda_n$ which satisfy

$$\lambda_1 \alpha_1 + \lambda_2 \alpha_2 + ..., + \lambda_n \alpha_n = \vec{0} ,$$ then the set of vectors are called linearly dependent .

4.8.Cofactor expansion of the determinant

Algebraic Cofactors of Determinant:

If A is a squae matrix ,then the minor of the entry in the i-th row and j-th colum (also called the $((i, j)$ minor) is the determinant of the submatrix formed by deleting the i-th row and j-th column .This number is often denoted $M_{i,j}$,The (i, j) cofactor is obtained by multiplying the minor by $(-1)^{i+j}$.

Cofactor expansion of the determinant:

Given the $n \times n$ matrix A ,the determinant of A can be written as the sum of the cofactors of any row or colum of the matrix multiplied by the entries that generated them .Namely, the cofactor expansion along the j-th colum gives :

$$\text{Det}(A)= a_{1j} M_{1j} + a_{2j} M_{2j} + ... + a_{nj} M_{nj}$$

The cofactor expansion along the ith row gives :

$$\text{Det}(A)= a_{i1} M_{i1} + a_{i2} M_{i2} + ... + a_{in} M_{in} .$$

4.9.Implicit Function Theorem

If binary function $F(x, y) = 0$ satisfies the following conditions :

(1) Equation $F(x_0, y_0) = 0$,

(2)Partial derivatives $F_x(x, y), F_y(x, y)$ are continuous in the neighbourhood

$D = \{(x, y) : (x - x_0)^2 + (y - y_0)^2 < \delta, \delta > 0\}$,

(3) $F_y(x_0, y_0) \neq 0$;

Then equation $F(x_0, y_0) = 0$ defines the only one continuous function $y = f(x)$,and the derived function

$$f'(x) = -\frac{F_x(x,y)}{F_y(x,y)}$$

is also continuous on the neighbourhood $\{x : (x - x_0)^2 < \varepsilon, \varepsilon > 0\}$.

4.10.Definition of Outer Product

Suppose $\alpha = (a_x, a_y, a_z)$, $\beta = (b_x, b_y, b_z)$ are two vectors in three-dimensional space .By using determinant calculation rules ,we define

$$\begin{vmatrix} i & j & k \\ a_x & a_y & a_z \\ b_x & b_y & b_z \end{vmatrix} = \begin{vmatrix} a_y & a_z \\ b_y & b_z \end{vmatrix} i - \begin{vmatrix} a_x & a_z \\ b_x & b_z \end{vmatrix} j + \begin{vmatrix} a_x & a_y \\ a_x & a_y \end{vmatrix} k = \left(\begin{vmatrix} a_y & a_z \\ b_y & b_z \end{vmatrix}, -\begin{vmatrix} a_x & a_z \\ b_x & b_z \end{vmatrix}, \begin{vmatrix} a_x & a_y \\ a_x & a_y \end{vmatrix} \right) = \gamma,$$

(where i, j, k are unit orthonormal vectors) ,then the vector γ is called to be the outer product of the vectors α, β ,or the vector product ,the cross product .

Vector γ is perpendicular to α, β ,and is perpendicular to $k_1\alpha + k_2\beta$, where k_1, k_2 are any real number .

5.Extremum of Multivariate Function under Equality Constraints

Before giving the new approaches to solving extremum of multivariate function under equality constraints ,we firstly review the interpretations for extreme value of function of one variable and the interpretation for binary function without constraint condition.

It is well known that a continuous function of one variable $y = f(x)$ expresses a curve .When the function is derivable and x_0 is the suspicious extreme point ,then $f'(x_0) = 0$ and the normal vector $N = (0,1)$,the tangent vector $N = (1,0)$ at the point x_0 .

It is also well known that binary function $z = f(x, y)$ expresses a surface in space and the normal vector in the face is $N = \pm(f_x, f_y, -1)$ .When the two partial derivatives $f_x(x, y), f_y(x, y)$ all exist and the $P(x_0, y_0)$ is the suspicious extreme point ,then $f_x(x_0, y_0) = 0, f_y(x_0, y_0) = 0$ and the normal vector is $N = \pm(0, 0, 1)$ at the point $P(x_0, y_0)$,and the tangent plane at the point P is perpendicular to Z-axis .

It is speculated from the above that normal vector ,tangent vector and tangent point probably can be the instruments to deal with extremum value of multivariate function under equality constraints.

5.1.Extreme value of binary function under constraints

5.1.1.A binary function is a family of curves and a moving curve in the plane

Currently, in textbooks (or reference books)of mathematics analysis, a binary function is only thought to be a surface in 3-dimensional space .In fact ,a binary function has many meanings.

Let me give a function $z = xy$. When z is assigned a non-zero real number, a hyperbola is obtained from the function ;when z has travered all the real number on the interval $(-\infty,0) \cup (0,+\infty)$,a family of hyperbola in the plane is obtained ; when z is changing continuously ,a hyperbola is changing accordingly in the plane .So,the function $z = xy$ denotes a family of hyperbola and a moving hyperbola in the plane .

From the perspective of the geometry , any continuous binary function expresses a family of curves and a moving curve in the plane ;and from the perspective of set theory ,any binary function expresses a mapping from the domain to the value region of the function ,which makes each curve and all the points on the each curve correspond to the only value of the function .There is no one-one correspondence from the points in 2-dimensional space to the points on the line .

But the family of curves determined by a binary function have at most common point.

5.1.2.Extreme value of binary function under equality constraint.

There ,generally known ,are three position relations between two curves :be apart from each other ,intersect each other ,be tangent to each other .If two curves are tangent at the point P ,there existes a commum tangent line and collinear normal vector at the point P .

Example 5 The equation $xy = t(x, y, t > 0)$ denotes a moving curve ,and the equation $(x-2)^2 + (y-2)^2 = 2$ denotes static circle .When $t = 1$ or 9 ,the moving curve is tangent to the static circle and the two tangent points are $(1,1)$ and $(3,3)$ respectively ,at which there are collinear normal $N = (1,1)$ and collinear tangent vector $v = (1,1)$;when $1 < t < 9$,the moving curve intersects the static circle ;when $0 < t < 1$ or $t > 9$,the moving curve and the static circle are apart away .If equation $xy = t$ is seen as a binary objective function $z = xy$ and $(x-2)^2 + (y-2)^2 = 2$ seen as an equality constraint ,then the tangent point $(1,1)$ is minimum point and $(3,3)$ is maximum point of the function under the constaint .

And by extension ,when the moving curve defined by the binary objective function is tangent to the statics curve defined by the constraint equation ,the tangent points are suspicious extreme points , at which there are commum tangent line and commum normal vector .This is the dynamic geometrical meaning of extreme value for binary fuction under equality constraint .

Because binary objective function denots a surface and equality constraint (namely binary equation) denots a cylinder in 3-dimensional space ,the combination of binary objective function and equality constraint expresses a intersecting line in space .Let a plane which is perpendicular to the z-axis be tangent to the intersecting line ,if the tangent point is (x_0, y_0, z_0), then the (x_0, y_0) is extreme point ,and z_0 is the extreme value .This is called static geometry meaning of binary function under equality constraint .

Of course ,it has economic significance and mechanical meaning .In view of the dynamic geometrical interpretation for extreme value of binary function under equality ,a theorem is derived .

Theorem 1 Objective function and equality constraint are as follows:

$$u = f(x, y), g(x, y) = 0.$$

Suppose the point $P(x_0, y_0)$ is suspicious extreme point ,then the two vectors

$$N = (f_x, f_y), N_0 = (g_x, g_y)$$

are collinear at (x_0, y_0) ,there exists non-zero real number λ which satisfies

$$(f_x, f_y)|_{(x_0, y_0)} + \lambda(g_x, g_y)|_{(x_0, y_0)} = (0,0).$$

Proof: If $P(x_0, y_0)$ is extreme point ,firstly $g(x_0, y_0) = 0$. Suppose the function $f(x, y), g(x, y)$ both have first continumous partial derivatives in the neighbourhood of the point P . Accordiong to theorem of implicit function ,the equation defines a function $y = y(x)$ which has first continuous order derivative. Plug $y = y(x)$ into objective function to chang it into the function of one variable $u = f[x, y(x)]$ .So the extreme point (x_0, y_0) where the objective function obtains extreme value is equivalent to the extreme point x_0 where the function $u = f[x, y(x)]$ obtains extreme value .

On account of the necessary condition by which function of one variabl obtains extreme value ,

$$\frac{du}{dx}|_{x=x_0} = f_x(x, y) + f_y(x, y)\frac{dy}{dx}|_{x=x_0} = 0;$$

According to theorem of implicit function ,

$$\frac{dy}{dx}|_{x=x_0} = -\frac{g_x(x_0, y_0)}{g_y(x_0, y_0)},$$

$$f_x(x_0, y_0) - f_y(x_0, y_0)\frac{g_x(x_0, y_0)}{g_y(x_0, y_0)} = 0,$$

Suppose $g_x(x, y), g_y(x, y) \neq 0$ at the point (x_0, y_0) ,then

$$\frac{f_x(x_0, y_0)}{g_x(x_0, y_0)} = \frac{f_y(x_0, y_0)}{g_y(x_0, y_0)};$$

Let

$$\frac{f_x(x_0, y_0)}{g_x(x_0, y_0)} = \frac{f_y(x_0, y_0)}{g_y(x_0, y_0)} = -\lambda,$$

then

$$f_x(x_0, y_0) = -\lambda f_g(x_0, y_0), f_y(x_0, y_0) = -\lambda g_y(x_0, y_0),$$

So,

$$(f_x, f_y)|_{(x_0, y_0)} + \lambda (g_x, g_y)|_{(x_0, y_0)} = (0,0).$$

Example 6 :Find out a point in line $xy = 1(x > 0)$ to make function $f(x, y) = x^2 + 2y^2$ obtain mimimum .

Solution :When $f(x, y) = x^2 + 2y^2$ obtains extreme value ,the following two vectors

$$N = (y, x), N_0 = (x, 2y)$$

are collinear , solve system of equations

$$\begin{cases} \dfrac{y}{x} = \dfrac{x}{2y} \\ xy = 1(x > 0) \end{cases}$$

Obtain the only exreme point $\left(\sqrt[4]{2}, \dfrac{1}{\sqrt[4]{2}} \right)$.

As the fuction $f(x, y) = x^2 + 2y^2$ is a moving ellipse , $xy = 1(x > 0)$ is a static curve , so exreme point $\left(\sqrt[4]{2}, \dfrac{1}{\sqrt[4]{2}} \right)$ namely , a tangent point is a mimimum point .

Example 7: Solve extreme points of function $u = x - 2y$ under the constraint

$$4x^2 - xy + 2y^2 = 465 .$$

Solution: The two normal vectors $N = (1, -2), N_0 = (8x - y, -x + 4y)$,

According to Theorem 1 ,when the extreme point is obtained ,the two normal vector are colliear, namely ,

$$\frac{8x - y}{1} = \frac{-x + 4y}{-2} ,$$

$$y = -\frac{15}{2} x ,$$

Solve the system of equations

$$\begin{cases} y = -\dfrac{15}{2} x \\ 4x^2 - xy + 2y^2 = 465 \end{cases} ,$$

We obtain extreme points $P_1(1, -\frac{15}{2}), P_2(-1, \frac{15}{2})$,corresponding extreme values are 16 and -16.

Because the domain defined by equality constraint is an ellipse and is a closed domain ,the value 16 is maximum and the value -16 is the minimum of the function.

It is difficult to solve the above question in Lagrange multiplier method .

5.1.2 Minimum and maximum of binary function under inequality constraints

While we are solving minimum and maximum of a variable function ,we must solve the values of stationary points ,the values of endpoints of closed interval ,then we can obtain minimum value and maximum value .Analogously ,while we are solving value domain of binary function under inequality constrains ,we must not only solve extreme values in domain but must also solve extreme values on the boundary curves and the values at the vertex in the domain .

Example 9 :Calculate mimimum and maxmimu of function $z = x^2 + y^2 - 12x + 16y$ in the domain

$D : x^2 + y^2 \leq 25$.

Solution :(1)First solve stationary point in the domain $x^2 + y^2 < 25$,

Let

$$\begin{cases} z_x = 2x - 12 = 0 \\ z_y = 2y + 16 = 0 \end{cases},$$

obtain point $(6, -8) \notin D$.

(2)Solve extreme points of the function under equality constraint $x^2 + y^2 = 25$,as the function obtain value ,vectors

$$N = (x - 6, 2y + 9), N_0 = (x, y)$$

are collinear ,solve the following system of equations

$$\begin{cases} \dfrac{x-6}{x} = \dfrac{y+9}{y} \\ x^2 + y^2 = 25 \end{cases},$$

Obtain extreme points $(3, -4), (-3, 4)$,calcualte

$$z(3, -4) = -75, z(-3, 4) = 125 \quad ,$$

Thereby $z_{\min} = -75, z_{\max} = 125$.

Example 10 Solve the value domain of the function $u = x + y^4$,the domain is formed by curves

$3x + 2y^2 = 9, x + y = 1$, positive x-axis and positive y-axis.

Solution:Let us record

$$l : u = x + y^4, l_1 : x + y = 1, l_2 : 3x + 2y^2 = 9, N = (1, 4y^3), N_1 = (1, 1), N_2 = (3, 4y).$$

Obviously ,the function $u = x + y^4$ has no extreme point in iner of the domain .The moving curve l is neither tangent to x-axis nor tangent to y-axis.

If l is tangent to l_1, the normal vectors N, N_1 are collinear, then

$$\frac{1}{1} = \frac{4y^3}{1},$$

We obtain $y = \dfrac{1}{\sqrt[3]{4}}$, substitute it into $l_1 : x + y = 1$,

we obtain the extreme point $P_1(1 - \dfrac{1}{\sqrt[3]{4}}, \dfrac{1}{\sqrt[3]{4}})$.

In a similar way, if l is tangent to l_2, then N, N_2 are collinear, namely

$$\frac{1}{3} = \frac{4y^3}{4y},$$

We obtain $y = \dfrac{1}{\sqrt{3}}$, then substitute it into $l_2 : 3x + 2y^2 = 9$,

We obtain the extreme point $P_2(\dfrac{25}{3}, \dfrac{1}{\sqrt{3}})$.

The vertex are $(0, \dfrac{3}{\sqrt{2}}), (3,0), (0,1), (1,0)$.

Calculate values :

$$u\big|_{(\frac{25}{3}, \frac{1}{\sqrt{3}})} = \frac{26}{9},\ u\big|_{(1-\frac{1}{\sqrt[3]{4}}, \frac{1}{\sqrt[3]{4}})} = 1 - \frac{3}{4 \cdot \sqrt[3]{4}},$$

$$u\big|_{(0, \frac{3}{\sqrt{2}})} = \frac{81}{4},\ u\big|_{(3,0)} = 3,\ u\big|_{(1,0)} = u\big|_{(0,1)} = 2.$$

Because the function is continuous in the domain, the value domain is the interval $[1 - \dfrac{3}{4 \cdot \sqrt[3]{4}}, \dfrac{81}{4}]$.

This method can be generalized to solve value domain of function of n-variables under inequality constraints .

5.2.Extremeum of ternary function under equality constraints

5.2.1.A ternary function is a family of surfaces and a moving surface in 3-dimensional space

A ternary equation with infinite many real solutions denots a static surface in 3-dimensional space .

Fistly, we are to study ternary function $u = x^2 + y^2 + z^2$. When u is assigned a different positive number ,a different sphere is obtained from the function ;we can realize that the function denotes a family of sphere and a moving sphere in 3-dimensional space .

In general ,any ternary function denotes a family of surfaces and a moving surface in 3-dimensional space .

From the perspective of the geometry , any continuous ternary function denotes a family of surfaces and a moving surface in the 3-dimensioal space ;and from the perspective of set theory ,any ternary function expresses a mapping from the domain to the value region of the function ,which makes each surface and all the points on the each surface correspond to the only value of the function .There is no one-one correspondence from the points in 3-dimensional space to the points on the line .

But the family of curves determined by a binary function have at most common point.

5.2.2.Extreme values of ternary function under equality constraint

There ,generally known ,are three position relations between two surfaces :be apart from each other ,intersect each other ,be tangent to each other .If two surfaces are tangent at the point P ,there existes a collinear normal vector at the point P . But ,the all commum tangent lines form a tangent plane at the tangent point .

Since any ternary function denotes a moving surface and any ternary equality denotes a static surface in the 3-dimensional space ,it is speculated that when the moving surface defined by the ternary objective function is tangent to the static surface defined by the constraint equation ,the tangent points are suspicious extreme points , at which there are collinear normal vector .This is the dynamic geometrical meaning of extreme value for ternary fuction under equality constraint .From what has been discussed the above ,we can consequently draw the following conclusions .

Theorem 2 Objective function and equality constraint are as follows:

$$u = f(x, y, z), g(x, y, z) = 0 .$$

Suppose the point $P(x_0, y_0, z_0)$ is suspicious extreme point ,then the two vectors

$$N = (f_x, f_y, f_z), N_0 = (g_x, g_y, g_z)$$

are collinear at the (x_0, y_0, z_0) ,there existes non-zero real number λ that satisfies

$$(f_x, f_y, f_z)|_{(x_0, y_0, z_0)} + \lambda(g_x, g_y, g_z)|_{(x_0, y_0, z_0)} = (0, 0, 0) .$$

Proof: If $P(x_0, y_0, z_0)$ is extreme point ,firstly $g(x_0, y_0, z_0) = 0$ .Suppose $f(x, y, z), g(x, y, z)$ both have first continuous partial derivatives .On the basis of theorem of implicit function , the equation $g(x, y, z) = 0$ defines a binary function $z = z(x, y)$ which has first continuous partial derivatives in some neighbourhood of the point (x_0, y_0).Then the objective function $u = f(x, y, z)$ is changed into the binary

function $u = f[x, y, z(x, y)]$ in the neighbourhood of the point (x_0, y_0);then the extreme point (x_0, y_0, z_0)

where the objective function obtains extreme value is equivalent to the extreme point (x_0, y_0) where the

function $u = f[x, y, z(x, y)]$ obtains extreme value .

Accorfing to the necessary condition by which binary function obtaines extreme value ,

$$\frac{\partial u}{\partial x}|_P = f_x(x, y, z) + f_z(x, y, z)\frac{\partial z}{\partial x}|_P = 0, \frac{\partial u}{\partial y}|_P = f_y(x, y, z) + f_z(x, y, z)\frac{\partial z}{\partial y}|_P = 0,$$

From the equation $g(x, y, z) = 0$,we can obtaine

$$\frac{\partial z}{\partial x}|_P = -\frac{g_x(x, y, z)}{g_z(x, y, z)}|_P, \frac{\partial z}{\partial y}|_P = -\frac{g_y(x, y, z)}{g_z(x, y, z)}|_P,$$

then

$$f_x(x, y, z)|_P = \frac{f_z(x, y, z)}{g_z(x, y, z)}g_x(x, y, z)|_P, f_y(x, y, z)|_P = \frac{f_z(x, y, z)}{g_z(x, y, z)}g_y(x, y, z)|_P,$$

Suppose $g_x(x, y, z), g_y(x, y, z), g_z(x, y, z) \neq 0$ at the point (x_0, y_0, z_0) ,then

$$\frac{f_x(x, y, z)}{g_x(x, y, z)}|_P = \frac{f_z(x, y, z)}{g_z(x, y, z)}|_P, \frac{f_y(x, y, z)}{g_y(x, y, z)}|_P = \frac{f_z(x, y, z)}{g_z(x, y, z)}|_P,$$

So ,

$$N = (f_x, f_y, f_z)|_P, N_0 = (g_x, g_y, g_z)|_P \text{ are collinear;}$$

Let

$$\frac{f_x(x, y, z)}{g_x(x, y, z)}|_P = \frac{f_y(x, y, z)}{g_y(x, y, z)}|_P = \frac{f_z(x, y, z)}{g_z(x, y, z)}|_P = -\lambda,$$

Then

$$(f_x, f_y, f_z)|_P + \lambda (g_x, g_y, g_z)|_P = (0, 0, 0).$$

Example 11: Solve extremem values of the function $u = x - 2y + 2z$ under the equality constrainte

$x^2 + y^2 + z^2 = 1$.

Solution:According to theorem 2 ,when the objective function obtaines extremem value ,the two

normal $N = (1, -2, 2), N_1 = (x, y, z)$ are collinear ,let

$$\frac{x}{1} = \frac{y}{-2} = \frac{z}{2} = t,$$

then $x = t, y = -2t, z = 2t$,

and then the extreme points are $\pm(\dfrac{1}{3}, -\dfrac{2}{3}, \dfrac{2}{3})$.

Let us deal with the question that we stated at the beginning.

Example 12: Slove the extreme point of the function $S = xy + 2(xz + yz)$ under the equality constraint

$xyz = V(x, y, z > 0)$.

Solution: According to theorem 2 ,when the objective function obtains extremem value ,the two normal

$N = (y + 2z, x + 2z, 2x + 2y)$, $N_1 = (yz, xz, xy)$.

are collinear ,namely

$$\frac{y + 2z}{yz} = \frac{x + 2z}{xz} = \frac{2x + 2y}{xy}$$

As $x, y, z \neq 0$,then

$$\frac{xy + 2xz}{xyz} = \frac{xy + 2yz}{xyz} = \frac{2xz + 2yz}{xyz},$$

$$xy + 2xz = xy + 2yz = 2xz + 2yz,$$

$$x = y = 2z,$$

Solve system of equations $\begin{cases} x = y = 2z \\ xyz = V \end{cases}$,

The only one extreme point is $\left(2\sqrt[3]{\dfrac{V}{4}}, 2\sqrt[3]{\dfrac{V}{4}}, \sqrt[3]{\dfrac{V}{4}} \right)$,it is the nimimum point too.

Example 13 : Solve the value domain of the function $u = xyz(x, y, z > 0)$,under the equality constraint

$ax + by + cz = 1(a, b, c > 0)$.

Solution: The domain is a triangle plane (including the three boundaries) in one octant .Let us firstly slove

extreme point ,according to theorem 2 ,when $u = xyz(x, y, z > 0)$ obtains extreme value ,the two normal

vectors

$N = (yz, xz, xy), N_1 = (a, b, c)$

are collinear ,namely

$$\frac{a}{yz} = \frac{b}{xz} = \frac{c}{xy},$$

$$\frac{ax}{xyz} = \frac{by}{xyz} = \frac{cz}{xyz},$$

$$ax = by = cz.$$

Solve the system of equations

$$\begin{cases} ax = by = cz \\ ax + by + cz = 1 \end{cases}$$

Obtaine the only one estreme point $P(\frac{1}{3a}, \frac{1}{3b}, \frac{1}{3c})$.

$$u_P = u\big|_{(\frac{1}{3a}, \frac{1}{3b}, \frac{1}{3c})} = \frac{1}{27abc}.$$

According to the geometry meaning of the objective function ,the value domain is interval $(0, \frac{1}{27abc}]$.

Example 14 Find out a point in the surface $x^2 + y^2 + \frac{z^2}{4} = 1(x, y, z > 0)$,at which there is a tangent plane ,to minimize the quadratic sum of the three interceptes ,at which the tangent plane intersects with x-axis ,with y-axis ,with z-axis .

Solution:Tangent plane at any point (x, y, z) in the surface is

$$xX + yY + \frac{zZ}{4} = 1,$$

where the point (x, y, z) is in the surface , point (X, Y, Z) is any point in tangent plane .

The interceptes of the tangent plane on three axes are $\frac{1}{x}, \frac{1}{y}, \frac{4}{z}$ respectively,

So the objective function is

$$u = \frac{1}{x^2} + \frac{1}{y^2} + \frac{16}{z^2}.$$

According to theorem 2 ,when the objective function obtaines extreme value ,the two normal vectors

$$N = (\frac{1}{x^3}, \frac{1}{y^3}, \frac{16}{z^3}), N_0 = (4x, 4y, z)$$

are collinear ,so

$$4x^4 = 4y^4 = \frac{z^4}{16} \quad \text{or} \quad x^2 = y^2 = \frac{z^2}{8}.$$

So the only extreme point is $(\frac{1}{2}, \frac{1}{2}, \sqrt{2})$.

5.2.2.Extreme values of ternary function under two equality

constraintes

A ternary eqution representes a surface .If two ternary equations have infinitely many commom solutions ,then the system of the two equations represents a intersecting line of them ,which direction vector(namely the tangent vector) at point on the intersecting line is the outer product of the two normal vectors at the cross point between the two planes .Direction vector is perpendicular to the two narmal vectors at the cross point .

In 3-dimensional space ,there are position relationships between a curve and a surface :be apart from each other ,be tangent to each oher and be interseceted with each other .If a surface is tangent to a curve ,then at the tangent point, the normal vector of the surface is perpendicular to the direction vector .In other words ,when the surface is tangent to intersecting line (defined by other surfaces) ,the three normal vectors of the three planes are coplanar at the tangent point .Two ternary equations with infinetly many solutions represent a static curve ,a ternary function representes a moving surface .Like the geometric meaning of extreme values of ternary function under equality constraint ,the extreme point of ternary function under two equality constraints is the point where the moving surface defined by objective function is tangent to the static curve defined by equality constrains , in the case ,the three normal vectors are coplanar at the tangent point .

Theorem 3 Objective function and equality constraints are as follows respectively

$$u = f(x,y,z), g(x,y,z) = 0, h(x,y,z) = 0,$$

denote

$$N = (\frac{\partial f}{\partial x}, \frac{\partial f}{\partial y}, \frac{\partial f}{\partial z}), N_1 = (\frac{\partial g}{\partial x}, \frac{\partial g}{\partial y}, \frac{\partial g}{\partial z}), N_2 = (\frac{\partial h}{\partial x}, \frac{\partial h}{\partial y}, \frac{\partial h}{\partial z}),$$

If $P(x_0, y_0, z_0)$ is suspicious extreme point ,then

$$(1) \begin{vmatrix} f_x & f_y & f_z \\ g_x & g_y & g_z \\ h_x & h_y & h_z \end{vmatrix}\Big|_{(x_0, y_0, z_0)} = 0,$$

(2)At the point $P(x_0, y_0, z_0)$,there exist the only two real numbers λ, μ not all zero ,such that

$$N + \lambda N_1 + \mu N_2 = (0,0,0).$$

Proof: (1)Suppose the curve defined by the constraint equations is

$$L : x = x(t), y = y(t), z = z(t),$$

Then the objective function is changed into unary function

$$u = f[x(t), y(t), z(t)].$$

Suppose the curve L is smooth ,then tangent vector to the curve is

$$(\frac{dx}{dt}, \frac{dy}{dt}, \frac{dz}{dt}) = \begin{vmatrix} i & j & k \\ g_x & g_y & g_z \\ h_x & h_y & h_z \end{vmatrix} = (\begin{vmatrix} g_y & g_z \\ h_y & h_z \end{vmatrix}, -\begin{vmatrix} g_x & g_z \\ h_x & h_z \end{vmatrix}, \begin{vmatrix} g_x & g_y \\ h_x & h_y \end{vmatrix}).$$

If the objective obtain extreme value at point $P(x_0, y_0, z_0)$,firstly

$$g(x_0, y_0, z_0) = 0 \, , h(x_0, y_0, z_0) = 0 \, , x_0 = x(t_0), y_0 = y(t_0), z_0 = z(t_0),$$

As such , $u = f[x(t), y(t), z(t)]$ obtains extreme value at $t = t_0$.

According to the necessary condition of by which unary function obtains extreme value

$$\frac{du}{dt}\Big|_{t_0} = f_x(x_0, y_0, z_0)\frac{dx}{dt}\Big|_{t_0} + f_y(x_0, y_0, z_0)\frac{dy}{dt}\Big|_{t_0} + f(x_0, y_0, z_0)\frac{dz}{dt}\Big|_{t_0} = 0,$$

namely

$$f_x(x, y, z)\begin{vmatrix} g_y & g_z \\ h_y & h_z \end{vmatrix}\Big|_{t_0} - f_y(x, y, z)\begin{vmatrix} g_x & g_z \\ h_x & h_z \end{vmatrix}\Big|_{t_0} + f_z\begin{vmatrix} g_x & g_y \\ h_x & h_y \end{vmatrix}\Big|_{t_0} = 0,$$

using determinantal property ,we obtain

$$\begin{vmatrix} f_x & f_y & f_z \\ g_x & g_y & g_z \\ h_x & h_y & h_z \end{vmatrix}\Big|_{(x_0, y_0, z_0)} = 0 .$$

(2)The above (1) shows that the three vectors are coplanar, then there exist only two real numbers λ, μ not all zero ,such that

$$N + \lambda N_1 + \mu N_2 = (0,0,0).$$

Example 15 :The paraboloid $z = x^2 + y^2$ intersects the plane $x + y + z = 1$ and forms an ellipse .Compute the maximum distance and minimum distance from points on the ellipse to original point . (From Pei Liwen *Typical Questions and Methods in Mathematical Analysis* 716:2006).

Solution :Suppose $P(x, y, z)$ is on the ellipse ,then the distance from points on the ellipse to the original point is

$$d = \sqrt{x^2 + y^2 + z^2} .$$

This question can be changed into the question of the extremem values of the objective function

$$u = x^2 + y^2 + z^2$$

under the equality constraints

$$x + y + z = 1, z = x^2 + y^2 .$$

Accoding to theorem 3 ,when the objective function obtains extreme value ,the three normal vectors are coplanar ,namely

$$0 = \begin{vmatrix} 1 & 1 & 1 \\ 2x & 2y & -1 \\ 2x & 2y & 2z \end{vmatrix} = (2z + 1)(y - x),$$

So we obtain two systems of equations

$$\begin{cases} 2z+1=0 \\ x+y+z=1 \\ x^2+y^2-z=0 \end{cases}, \begin{cases} x=y \\ x+y+z=1 \\ x^2+y^2-z=0 \end{cases},$$

The first system of equations has no solution ,from the second system ,we obtain extremem points.

$$P(\frac{-1-\sqrt{3}}{2},\frac{-1-\sqrt{3}}{2},2+\sqrt{3}),Q(\frac{-1+\sqrt{3}}{2},\frac{-1+\sqrt{3}}{2},2-\sqrt{3});$$

Then we obtain

$$u\mid_P=9+5\sqrt{3},u\mid_Q=9-5\sqrt{3}.$$

Because the objective function is continuous in the closed domain ,maximum distance and minimum distance are $\sqrt{9+5\sqrt{3}},\sqrt{9-5\sqrt{3}}$ respectively .

Example 16 The ellipsoid $\frac{x^2}{3}+y^2+\frac{z^2}{2}=1$ intersects the plane $2x+y+z=0$ and forms an ellipse .Compute the area of the ellipse .

Solution:We can compute the area of the ellipse as along as we compute the major axis and minor axis of the ellipse ,so the question is translated into the question to computing the extremem value of the function $r^2=x^2+y^2+z^2$ under the equality constraints

$$\frac{x^2}{3}+y^2+\frac{z^2}{2}=1,2x+y+z=0.$$

Accoding to theorem 3,when the objective function obtains extreme value ,then the three normal vectors are coplanar ,namely

$$0=\begin{vmatrix} 2 & 1 & 1 \\ \frac{x}{3} & y & \frac{z}{2} \\ x & y & z \end{vmatrix}=\begin{vmatrix} 2 & 1 & 1 \\ 0 & 4y & z \\ 0 & -x+2y & -x+2z \end{vmatrix}=-8xy+12yz+2xz ,$$

From the following system of equations

$$\begin{cases} 2x+y+z=0 \\ -8xy+12yz+2xz=0 \end{cases}$$

We obtain

$$4(\frac{y}{z})^2+15(\frac{y}{z})-1=0 ,$$

$$\frac{y}{z}=\frac{-15\pm\sqrt{241}}{8} .$$

Because $x,y,z\neq 0$,

$$0 = \begin{vmatrix} 2 & 1 & 1 \\ \dfrac{x}{3} & y & \dfrac{z}{2} \\ x & y & z \end{vmatrix} = \begin{vmatrix} 2x & y & z \\ \dfrac{x^2}{3} & y^2 & \dfrac{z^2}{2} \\ x^2 & y^2 & z^2 \end{vmatrix} = \begin{vmatrix} 2x+y+z & y & z \\ \dfrac{x^2}{3}+y^2+\dfrac{z^2}{2} & y^2 & \dfrac{z^2}{2} \\ x^2+y^2+z^2 & y^2 & z^2 \end{vmatrix} = \begin{vmatrix} 0 & y & z \\ 1 & y^2 & \dfrac{z^2}{2} \\ r^2 & y^2 & z^2 \end{vmatrix}$$

$$= \begin{vmatrix} 0 & 1 & 1 \\ 1 & y & \dfrac{z}{2} \\ r^2 & y & z \end{vmatrix} = \begin{vmatrix} 0 & 0 & 1 \\ 1 & y-\dfrac{z}{2} & \dfrac{z}{2} \\ r^2 & y-z & z \end{vmatrix} = y - z - r^2(y - \dfrac{z}{2}).$$

Then we obtain

$$r^2 - 1 = \frac{z}{z - 2y} = \frac{1}{1 - 2\dfrac{y}{z}} = \frac{1}{1 - 2\dfrac{-15 \pm \sqrt{241}}{8}} = \frac{4}{19 \pm \sqrt{241}} = \frac{19 \pm \sqrt{241}}{30},$$

$$r_1^2 = \frac{49 + \sqrt{241}}{30}, r_2^2 = \frac{49 - \sqrt{241}}{30}; r_1^2 r_2^2 = \frac{49^2 - 241}{900} = \frac{12}{5};$$

Because r_1, r_2 are the maximum and minimum of the function ,also are the major axis and minor axis of the ellipse respectively ,so the area is

$$S = \pi r_1 r_2 = 2\pi\sqrt{\frac{3}{5}}.$$

Example 17 Objective function and constrain conditions are as follow

$$u = ax^2 + by^2 + cz^2 (a, b, c >) \quad , \quad x + y + z = 1 (x, y, z \geq 0),$$

calculate the maximum value and minimum value of the function .

Solution:The ojective function denotes a moving ellipsoid and the constraints denote a domain of a equilateral triangle (incuding the three sides of it) . While the moving ellipsoid is increasing from a point ,the position relationship between the moving ellipsoid and the domain is apart from ,tangent , intersectant ,apart from once again .We firstly compute extreme point of the function (namely tangent point) .According to theorem 2 ,when the function obtaines extreme value ,

$$\frac{2ax}{1} = \frac{2by}{1} = \frac{2cz}{1},$$

So ,solve the following system of equations

$$\begin{cases} ax = by = cz \\ x + y + z = 1 \end{cases}$$

We can obtain the only one extreme point

$$P(\frac{bc}{ab + bc + ca}, \frac{ca}{ab + bc + ca}, \frac{ab}{ab + bc + ca}),$$

From geometric meaning ,the only extreme point is minimum point ,

$$u_{\min} = \frac{abc}{ab + bc + ca} \quad ,$$

Compute the values at the vertex of the triangle and compare them

$$u\big|_{(1,0,0)} = a, u\big|_{(0,1,0)} = b, u\big|_{(0,0,1)} = c ,$$

So $u_{\min} = \dfrac{abc}{ab+bc+ca}, u_{\max} = \max\{a,b,c\}$.

The question is the extension for a question in the postgraduate entrance examination of Fudan University in 1999 year .

5.3.Extremem values of function of n variables under one equality constrain or under n-1equality constraints (n≥4)

5.3.1.Meanings of function of n variables

In order to deduce the method of solving function of n variables under one equality constrain or under n-1equality constraints (n≥4) ,we must generalize surface in 3-dimensional to surface in n-dimensional and give geometric meanings of it .

Definition 3   Function $u = f(x_1, x_2, ..., x_n)$ denotes a moving surface and a family of surfaces in n-dimensional ,which normal vector at point in the surface is $N = (f_{x_1}, f_{x_2}, ..., f_{x_n})$.

5.3.2.Extremem values of function of n variables under one equality constraint

Theorem 4   The objective function of n variables and equality contraint (n≥2) are as follows

$$u = f(x_1, x_2, ..., x_n), F(x_1, x_2, ..., x_n) = 0 .$$

Suppose the point $P_0(x_1, x_2, ..., x_n)$ is the suspicious point of the objective function ,then

(1) $\dfrac{f_{x_1}}{F_{x_1}} = \dfrac{f_{x_2}}{F_{x_2}} = ... = \dfrac{f_{x_n}}{F_{x_n}}$,

(2)there exists a real number λ ,so that

$$(f_{x_1}, f_{x_2}, ..., f_{x_n}) + \lambda(F_{x_1}, F_{x_2}, ..., F_{x_n}) = \vec{0} .$$

The proof of theorem 4 is simlar to the proofs of theorem 1 and theorem 2 .

Example 18: The objective function and equality contraint are as follows respectively

$$u = x_1^2 + x_2^2 + ... + x_n^2 \ , \ \frac{x_1}{a_1} + \frac{x_2}{a_2} + ... + \frac{x_n}{a_n} = 1 \ ,$$

where the constants $a_i > 0 (i = 1,2,...,n)$ and the variables $x_i > 0 (i = 1,2,...,n)$ .Solve extreme value of the objective function .

Solution :According to Theorem 4 ,if the function obtains extreme value ,the normal vectors

$$N_0 = (\frac{1}{a_1}, \frac{1}{a_2}, ..., \frac{1}{a_n}), N = (x_1, x_2, ..., x_n)$$

are collinear ,so let

$$a_1 x_1 = a_2 x_2 = ... = a_n x_n = t,$$

obtain

$$x_1 = \frac{t}{a_1}, x_2 = \frac{t}{a_2}, ..., x_n = \frac{t}{a_n},$$

Plug them into constraint equation ,obtaint

$$t = \frac{1}{\frac{1}{a_1^2} + \frac{1}{a_2^2} + ... + \frac{1}{a_n^2}},$$

and the only one extreme point ,whose coordinates are

$$x_1 = \frac{1}{a_1(\frac{1}{a_1^2} + \frac{1}{a_2^2} + ... + \frac{1}{a_n^2})}, x_2 = \frac{1}{a_2(\frac{1}{a_1^2} + \frac{1}{a_2^2} + ... + \frac{1}{a_n^2})}, ..., x_n = \frac{1}{a_n(\frac{1}{a_1^2} + \frac{1}{a_2^2} + ... + \frac{1}{a_n^2})}.$$

From perspective of geometry ,the constraint equation denotes a plane in n-dimensional space and objective function denotes a moving sphere in the same space ,so the extremem point is minimum value .

$$u_{\min} = x_1^2 + x_2^2 + ... + x_n^2$$

$$= (\frac{1}{a_1^2} + \frac{1}{a_2^2} + ..., + \frac{1}{a_n^2})t^2$$

$$= (\frac{1}{a_1^2} + \frac{1}{a_2^2} + ... + \frac{1}{a_n^2})(\frac{1}{\frac{1}{a_1^2} + \frac{1}{a_2^2} + ... + \frac{1}{a_n^2}})^2$$

$$= \frac{1}{\frac{1}{a_1^2} + \frac{1}{a_2^2} + ... + \frac{1}{a_n^2}}.$$

Example 19: Objective function and equality constraint are as follows respectively

$$u = \frac{b_1}{x_1} + \frac{b_2}{x_2} + ... + \frac{b_n}{x_n} \quad , \quad a_1 x_1 + a_2 x_2 + ... + a_n x_n = 1;$$

where constants $a_1, a_2, ..., a_n, b_1, b_2, ..., b_n$ and variables $x_1, x_2, ..., x_n$ are all positive .Calculate extreme value of the objective function .

Solution: According to Theorem 4 ,if the function obtains extreme value ,the normal vectors

$$N_0 = (a_1, a_2, ...a_n), N = -(\frac{b_1}{x_1^2}, \frac{b_2}{x_2^2}, ..., \frac{b_n}{x_n^2})$$

are collinear ,so let

$$\frac{a_1 x_1^2}{b_1} = \frac{a_2 x_2^2}{b_2} = ... = \frac{a_n x_n^2}{b_n} = t ,$$

Then obtain

$$x_1 = \sqrt{\frac{b_1 t}{a_1}}, x_2 = \sqrt{\frac{b_2 t}{a_2}}, ..., x_n = \sqrt{\frac{b_n t}{a_n}} ,$$

$$\frac{1}{x_1} = \sqrt{\frac{a_1}{b_1 t}}, \frac{1}{x_2} = \sqrt{\frac{a_2}{b_2 t}}, ..., \frac{1}{x_n} = \sqrt{\frac{a_n}{b_n t}} ,$$

Plug them into constraint equation

$$a_1 \sqrt{\frac{b_1 t}{a_1}} + a_2 \sqrt{\frac{b_2 t}{a_2}} + ... + a_n \sqrt{\frac{b_n t}{a_n}} = 1 ,$$

$$(\sqrt{a_1 b_1} + \sqrt{a_2 b_2} + ... + \sqrt{a_n b_n})\sqrt{t} = 1 ,$$

We obtain

$$\frac{1}{\sqrt{t}} = \sqrt{a_1 b_1} + \sqrt{a_2 b_2} + ... + \sqrt{a_n b_n} ,$$

According to geometric meaning of this question ,the function has no maximum ,has minimum

$$u_{\min} = b_1 \sqrt{\frac{a_1}{b_1 t}} + b_2 \sqrt{\frac{a_2}{b_2 t}} + ..., + b_n \sqrt{\frac{a_n}{b_n t}}$$

$$= (\sqrt{a_1 b_1} + \sqrt{a_2 b_2} + ... + \sqrt{a_n b_n})\frac{1}{\sqrt{t}}$$

$$= (\sqrt{a_1 b_1} + \sqrt{a_2 b_2} + ... + \sqrt{a_n b_n})^2 .$$

5.3.3.Extreme value of function of n variables under n-1 equality constraints .

In order to give the approach to solving exreme value for function of n variables under n-1 equality constraints ,we must introduce the following conclusion to define curve in n-dimensional space and it's direction vector ,

Lemma 1 Suppose $\alpha_i = (a_{i1}, a_{i2}, ..., a_{in})(i = 1, 2, ..., n-1)$ are n-1 real vectors in n-dimensional space .By using determinant calculation rules,we define the operation

$$\beta = \begin{vmatrix} j_1 & j_2 & \cdots & j_n \\ a_{11} & a_{12} & \cdots & a_{1n} \\ \cdots & \cdots & \cdots & \cdots \\ a_{n-1,1} & a_{n-1,2} & \cdots & a_{n-1,n} \end{vmatrix} = A_1 j_1 + A_2 j_2 + ... + A_n j_n = (A_1, A_2, ..., A_n) ,$$

(where $j_1, j_2,..., j_n$ are unit orthonormal vectors , $A_1, A_2,..., A_n$ are corresponding cofactors). Then the vector β is called to be the outer product of the vectors $\alpha_1, \alpha_2,.., \alpha_{n-1}$,and $\beta = (A_1, A_2,..., A_n)$ is perpendicular to $\alpha_1, \alpha_2,.., \alpha_{n-1}$.

Note :In order to express concisely , the outer product of the n-1 vectors in n-dimensional space is expressed as $\beta = \{\alpha_1, \alpha_2,..., \alpha_{n-1}\}^{\perp}$.

From Lemma 1 ,one can easily argue that if the vector $\beta = \{\alpha_1, \alpha_2,..., \alpha_{n-1}\}^{\perp}$, then β is perpendicular to $k_1\alpha_1 + k_2\alpha_2 + ... + k_{n-1}\alpha_{n-1}$, where $k_1, k_2,..., k_{n-1}$ are any real numbers .

Please refer to *Generaliztion of Outer Product and Application* for more details about outer product .

Since two ternary eqution (which have infinitely many comom solutions)denotes a curve ,we can define curve in n-dimensional space .

Definition 5 If the following system of linear equations

$$\begin{cases} a_{11}x + a_{12}x + ... + a_{1n}x = b_1 \\ a_{21}x + a_{22}x + ... + a_{2n}x = b_2 \\ ... \\ a_{n-1,1}x + a_{n-1,2}x + ... + a_{n-1,n}x = b_{n-1} \end{cases}$$

has infinetely many solutions ,and the rank of the coefficient matrix is n-1 ,then the system of linear equatins denotes a straight line in n-dimensional space .The system of linear equations is called to be the general equation of straight line in n-dimensional ,which direction vector is

$$v = \begin{vmatrix} i_1 & i_2 & ... & i_n \\ a_{11} & a_{12} & ... & a_{1n} \\ ... & ... & ... & ... \\ a_{n-1,1} & a_{n-1,2} & ... & a_{n-1,n} \end{vmatrix} = (A_1, A_2,..., A_n) ,$$

where $A_1, A_2,..., A_n$ are corresponding to $i_1, i_2,..., i_n$ respectively.

Definetion 6 If the following system of equations

$$F_i(x_1, x_2,..., x_n) = 0, i = 1,2,..., n-1 ,$$

has infinitely many solutions ,and at any point of the solutions the rank of the Jacobian matrix

$$\frac{\partial(F_1, F_2,..., F_{n-1})}{\partial(x_1, x_2,..., x_n)} = \begin{bmatrix} \dfrac{\partial F_1}{\partial x_1} & \dfrac{\partial F_1}{\partial x_2} & ... & \dfrac{\partial F_1}{\partial x_n} \\ \dfrac{\partial F_2}{\partial x_1} & \dfrac{\partial F_2}{\partial x_2} & ... & \dfrac{\partial F_2}{\partial x_n} \\ ... & ... & ... & ... \\ \dfrac{\partial F_{n-1}}{\partial x_1} & \dfrac{\partial F_{n-1}}{\partial x_2} & ... & \dfrac{\partial F_{n-1}}{\partial x_n} \end{bmatrix}$$

is equal to n-1 ,then the system of equations denotes a curve in n-dimensional space ,whose tangent vector at point on the curve in n-dimensional space is

$$v = \begin{vmatrix} i_1 & i_2 & \cdots & i_n \\ \dfrac{\partial F_1}{\partial x_1} & \dfrac{\partial F_1}{\partial x_2} & \cdots & \dfrac{\partial F_1}{\partial x_n} \\ \cdots & \cdots & \cdots & \cdots \\ \dfrac{\partial F_{n-1}}{\partial x_1} & \dfrac{\partial F_{n-1}}{\partial x_2} & \cdots & \dfrac{\partial F_{n-1}}{\partial x_n} \end{vmatrix} = (A_1, A_2, ..., A_n) .$$

After the above definitions of curve in n-dimensioanl space and direction vector of the curve ,we generalize the extreme value for ternary function under equlity constraints ,we can arrive at the following conclusion .

Theorem 5 suppose the objective function of n-variables (n$\geq$2) and equality constraints are as follow respectively

$$u = f(x_1, x_2, ..., x_n), g_i(x_1, x_2, ..., x_n) = 0 (i = 1, 2, ..., n-1)$$

If the point P **is** the suspicious extreme point of the objective function ,then at the point P
(1) the following determinant

$$\begin{vmatrix} \dfrac{\partial f}{\partial x_1}\Big|_P & \dfrac{\partial f}{\partial x_2}\Big|_P & \cdots & \dfrac{\partial f}{\partial x_n}\Big|_P \\ \dfrac{\partial g_1}{\partial x_1}\Big|_P & \dfrac{\partial g_2}{\partial x_2}\Big|_P & \cdots & \dfrac{\partial f}{\partial x_n}\Big|_P \\ \cdots & \cdots & \cdots & \cdots \\ \dfrac{\partial g_{n-1}}{\partial x_1}\Big|_P & \dfrac{\partial g_{n-1}}{\partial x_2}\Big|_P & \cdots & \dfrac{\partial g_{n-1}}{\partial x_n}\Big|_P \end{vmatrix} = 0 ;$$

(2) there exist n-1 real numbers $\lambda_1, \lambda_2, ..., \lambda_{n-1}$ which satisfay

$$N + \lambda_1 N_1 + \lambda_2 N_2 + ... + \lambda_{n-1} N_{n-1} = \vec{0} ,$$

where $N, N_1, N_2, ..., N_{n-1}$ represent the normal vectors of surfaces in n-dimensioanl space defined by objective function and constraint equations respectively .

The proof of this theorem is similar to the proof of the theorem 3 .Here the system of n-1 constraint equations represents a satic curve ,the objective function represents a moving surface .When the objective function obtains extreme value ,the moving surface is tangent to the satic curve ,so that the extreme point is the tangent point ,then the n normal vectors of the n surfaces defined by function and constraint equations are linearly dependent .

According to theorem 6 ,when we solve the extreme value for function of n-variables under $n-1$ equality constraints ,we can use n normal vectors to establish a n-th order determinant to obtain an equality of n-variables ,thus from a system of equations (n$\times$n) ,we can obtain extreme points .This also applies to binary function under an equality constraint .

5.4. Extreme value for function of n-variables under m quality constraints (m<n)

Two normal vectors are collinear ,three normal vectors are coplanar and the determinant constructed by n normal vectors in n-dimensioanl space is zero .All of these cases are the special ones that normal vectors are linearly dependent .According to Theorem 3 and Theorem 6 , by futher generalizing the above cases ,it is easily seen to solve extreme points of multivariate function under constraints is related to determinant ,thereby the following conclusion can be speculated .

Theorem 6 : Suppose the objective function of n-variables and equality constraints are as follows respectively

$$u = f(x_1, x_2, ..., x_n) , F_i(x_1, x_2, ..., x_n) = 0 (i = 1, 2, ..., m; m < n) .$$

The normal vectors of surfaces defined by the function and constraint equations are expressed as

$$N = (\frac{\partial f}{\partial x_1}, \frac{\partial f}{\partial x_2}, ..., \frac{\partial f}{\partial x_n}), \ N_i = (\frac{\partial F_i}{\partial x_1}, \frac{\partial F_i}{\partial x_2}, ..., \frac{\partial f_i}{\partial x_n})(i = 1, 2, ..., m) .$$

If the point $P(x_1^0, x_2^0, ..., x_n^0)$ is suspicious extreme point of the function ,then

(1)there exsit real numbers $\lambda_1, \lambda_2, ..., \lambda_m$ which satisfy $N \mid_P + \sum_1^m \lambda_i N_i \mid_P = \vec{0}$,

(2) the all subdeterminants (m+1) $\times$ (m+1) of the matrix $E_{(m+1) \times n}$ fomed by row vectors

$N, N_1, N_2, ..., N_m$ are zero .

In order to prove Theorem 6 ,we must introduce the following lemma:

Lemma 2 If system of functions of m+n variables

$$F_i(x_1, x_2, ..., x_n, y_1, y_2, ..., y_m), i = 1, 2, ..., m$$

satisfy the following conditions at the point $P_0(x_1^0, x_2^0, ..., x_n^0, y_1^0, y_2^0, ..., y_m^0)$:

(1)value of every function is zero at the point P_0 ;

(2)every function has first continuous partial derivatives in some neighbourhood $\{P : \mid PP_0 \mid < \delta, \delta > 0\}$;

(3) Jacobian determinant

$$\mid \frac{\partial(F_1, F_2, ..., F_m)}{\partial(y_1, y_2, ..., y_m)} \mid \neq 0 \ ,$$

at the point P_0 ;Then :

(1)the system of equtions $F_i(x_1, x_2, ..., x_n, y_1, y_2, ..., y_m) = 0, i = 1, 2, ..., m$ define the only system of implicit

functions $y_i = f_i(x_1, x_2, ..., x_m), i = 1, 2, ..., m$ in some neighbourhood $\{P : \mid PP_0 \mid < \delta, \delta > 0\}$,and the system of implicit functions satisfy the system of equations

$$F_i[x_1, x_2, ..., x_n, f_1(x_1, x_2, ..., x_n), \ f_2(x_1, x_2, ..., x_n), ..., f_m(x_1, x_2, ..., x_m)] = 0,$$

and $y_i^0 = f_i(x_1^0, x_2^0, ..., x_n^0), i = 1, 2, ..., m$.

(2)in certain neighbourhood of the point $P_0'(x_1^0, x_2^0, ..., x_n^0)$,the system of implicit functions

$$y_i = f_i(x_1, x_2, ..., x_m), i = 1, 2, ..., m$$

have first continuous partial derivatives and

$$\begin{bmatrix} \dfrac{\partial y_1}{\partial x_1} & \dfrac{\partial y_1}{\partial x_2} & \cdots & \dfrac{\partial y_1}{\partial x_n} \\ \dfrac{\partial y_2}{\partial x_1} & \dfrac{\partial y_2}{\partial x_2} & \cdots & \dfrac{\partial y_2}{\partial x_n} \\ \cdots & \cdots & \cdots & \cdots \\ \dfrac{\partial y_m}{\partial x_1} & \dfrac{\partial y_m}{\partial x_2} & \cdots & \dfrac{\partial y_m}{\partial x_n} \end{bmatrix} = - \begin{bmatrix} \dfrac{\partial F_1}{\partial y_1} & \dfrac{\partial F_1}{\partial y_2} & \cdots & \dfrac{\partial F_1}{\partial y_m} \\ \dfrac{\partial F_2}{\partial y_1} & \dfrac{\partial F_2}{\partial y_2} & \cdots & \dfrac{\partial F_2}{\partial y_m} \\ \cdots & \cdots & \cdots & \cdots \\ \dfrac{\partial F_m}{\partial y_1} & \dfrac{\partial F_m}{\partial y_2} & \cdots & \dfrac{\partial F_m}{\partial y_m} \end{bmatrix}^{-1} \begin{bmatrix} \dfrac{\partial F_1}{\partial x_1} & \dfrac{\partial F_1}{\partial x_2} & \cdots & \dfrac{\partial F_1}{\partial x_n} \\ \dfrac{\partial F_2}{\partial x_1} & \dfrac{\partial F_2}{\partial x_2} & \cdots & \dfrac{\partial F_2}{\partial x_n} \\ \cdots & \cdots & \cdots & \cdots \\ \dfrac{\partial F_m}{\partial x_1} & \dfrac{\partial F_m}{\partial x_2} & \cdots & \dfrac{\partial F_m}{\partial x_n} \end{bmatrix}.$$

For the convenience of using Lemma 2 to prove theorem 6 ,we must give the following equivalent theorem of Theorem 6 .

Theorem 6.1 Suppsoe the objective function and equality constraints are as follows respectively

$$u = f(x_1, x_2, ..., x_n, y_1, y_2, ..., y_m) , F_i(x_1, x_2, ..., x_n, y_1, y_2, ..., y_m) = 0, i = 1,2,...,m ,$$

Let $N = (\dfrac{\partial f}{\partial x_1}, \dfrac{\partial f}{\partial x_2}, ..., \dfrac{\partial f}{\partial x_n}, \dfrac{\partial f}{\partial y_1}, ..., \dfrac{\partial f}{\partial y_m}), N_i = (\dfrac{\partial F_i}{\partial x_1}, \dfrac{\partial F_i}{\partial x_2}, ..., \dfrac{\partial F_i}{\partial x_n}, \dfrac{\partial F_i}{\partial y_1}, ..., \dfrac{\partial F_i}{\partial y_1}), i = 1,2,...,m \cdot$

If the point $P(x_1^0, x_2^0, ..., x_n^0, y_1^0, y_2^0, ..., y_m^0)$ is the suspicious point of the objective function ,then

(1)at the point ,the vectors $N, N_1, N_2, ..., N_m$ are linearly dependent .

(2) the all subdeterminants (m+1)$\times$(m+1) of the following matrix $E_{(m+1)(n+m)}$ are zero

$$E = \begin{bmatrix} \dfrac{\partial f}{\partial x_1} & \dfrac{\partial f}{\partial x_2} & \cdots & \dfrac{\partial f}{\partial x_n} & \dfrac{\partial f}{\partial y_1} & \cdots & \dfrac{\partial f}{y_m} \\ \dfrac{\partial F_1}{\partial x_1} & \dfrac{\partial F_1}{\partial x_2} & \cdots & \dfrac{\partial F_1}{\partial x_n} & \dfrac{\partial F_1}{\partial y_1} & \cdots & \dfrac{\partial F_1}{\partial y_m} \\ \cdots & \cdots & \cdots & \cdots & \cdots & \cdots & \cdots \\ \dfrac{\partial F_m}{\partial x_1} & \dfrac{\partial F_m}{\partial x_2} & \cdots & \dfrac{\partial F_m}{\partial x_n} & \dfrac{\partial F_m}{\partial y_1} & \cdots & \dfrac{\partial F_m}{\partial y_m} \end{bmatrix}$$

Proof :(1)Suppose at $P(x_1^0, x_2^0, ..., x_n^0, y_1^0, y_2^0, ..., y_m^0)$, the constraint system of equations satisfy the conditions of Lemma 2 ,then the constraint system of equations define the only system of impicit functions

$$y_i = y_i(x_1, x_2, ..., x_m), i = 1,2,...,m ,$$

in some neighbourhood of the point P ,and the system of implicit functions have first continuous partial derivatives ,which makes the objective function of m+n variables be converted into the function of n variables without any constraints

$$u = f[x_1, x_2, ..., x_n, y_1(x_1, x_2, ..., x_n), y_2(x_1, x_2, ..., x_n), ..., ..., y_m(x_1, x_2, ..., x_n)]$$

in some neighbourhood of the point $P'(x_1^0, x_2^0, ..., x_n^0)$.

According to necessary condition by which function of n variables without any constraints obtains extreme value ,we take partial derivative of functions f, F_i $(i = 1,2,...,m)$ with respect to x_1 ,and then let them be zero ,we can acquire the following equations of partial derivatives

$$\dfrac{\partial f}{\partial x_1} = -\dfrac{\partial f}{\partial y_1}\dfrac{\partial y_1}{\partial x_1} - \dfrac{\partial f}{\partial y_2}\dfrac{\partial y_2}{\partial x_1} - ... - \dfrac{\partial f}{\partial y_m}\dfrac{\partial y_m}{\partial x_1} ,$$

$$\frac{\partial F_1}{\partial x_1} = -\frac{\partial F_1}{\partial y_1}\frac{\partial y_1}{\partial x_1} - \cdots - \frac{\partial F_1}{\partial y_m}\frac{\partial y_m}{\partial x_1}$$

$$\frac{\partial F_2}{\partial x_1} = -\frac{\partial F_2}{\partial y_1}\frac{\partial y_1}{\partial x_1} - \cdots - \frac{\partial F_2}{\partial y_m}\frac{\partial y_m}{\partial x_1}$$

...

$$\frac{\partial F_m}{\partial x_1} = -\frac{\partial F_m}{\partial y_1}\frac{\partial y_1}{\partial x_1} - \cdots - \frac{\partial F_m}{\partial y_m}\frac{\partial y_m}{\partial x_1};$$

Convert the m+1 equations of partial derivatives into the following matrix eqution

$$\begin{bmatrix}\dfrac{\partial f}{\partial x_1}\\[4pt]\dfrac{\partial F_1}{\partial x_1}\\[4pt]\cdots\\[4pt]\dfrac{\partial F_m}{\partial x_1}\end{bmatrix} = -\begin{bmatrix}\dfrac{\partial f}{\partial y_1}&\dfrac{\partial f}{\partial y_2}&\cdots&\dfrac{\partial f}{\partial y_m}\\[4pt]\dfrac{\partial F_1}{\partial y_1}&\dfrac{\partial F_1}{\partial y_2}&\cdots&\dfrac{\partial F_1}{\partial y_m}\\[4pt]\cdots&\cdots&\cdots&\cdots\\[4pt]\dfrac{\partial F_m}{\partial y_1}&\dfrac{\partial F_m}{\partial y_2}&\cdots&\dfrac{\partial F_m}{\partial y_m}\end{bmatrix}\begin{bmatrix}\dfrac{\partial y_1}{\partial x_1}\\[4pt]\dfrac{y_2}{\partial x_1}\\[4pt]\cdots\\[4pt]\dfrac{\partial y_m}{\partial x_1}\end{bmatrix},$$

In the same way , we take partial derivative of functions f, F_i ($i = 1,2,...,m$) with respect to x_2 ,and convert m+1 equations of partial derivatives into following the matrix eqution

$$\begin{bmatrix}\dfrac{\partial f}{\partial x_2}\\[4pt]\dfrac{\partial F_1}{\partial x_2}\\[4pt]\cdots\\[4pt]\dfrac{\partial F_m}{\partial x_2}\end{bmatrix} = -\begin{bmatrix}\dfrac{\partial f}{\partial y_1}&\dfrac{\partial f}{\partial y_2}&\cdots&\dfrac{\partial f}{\partial y_m}\\[4pt]\dfrac{\partial F_1}{\partial y_1}&\dfrac{\partial F_1}{\partial y_2}&\cdots&\dfrac{\partial F_1}{\partial y_m}\\[4pt]\cdots&\cdots&\cdots&\cdots\\[4pt]\dfrac{\partial F_m}{\partial y_1}&\dfrac{\partial F_m}{\partial y_2}&\cdots&\dfrac{\partial F_m}{\partial y_m}\end{bmatrix}\begin{bmatrix}\dfrac{\partial y_1}{\partial x_2}\\[4pt]\dfrac{y_2}{\partial x_2}\\[4pt]\cdots\\[4pt]\dfrac{\partial y_m}{\partial x_2}\end{bmatrix},$$

...

We take partial derivative of functions f, F_i ($i = 1,2,...,m$) with respect to x_n ,and convert m+1 equations of partial derivatives into following the matrix eqution

$$\begin{bmatrix}\dfrac{\partial f}{\partial x_n}\\[4pt]\dfrac{\partial F_1}{\partial x_n}\\[4pt]\cdots\\[4pt]\dfrac{\partial F_m}{\partial x_n}\end{bmatrix} = -\begin{bmatrix}\dfrac{\partial f}{\partial y_1}&\dfrac{\partial f}{\partial y_2}&\cdots&\dfrac{\partial f}{\partial y_m}\\[4pt]\dfrac{\partial F_1}{\partial y_1}&\dfrac{\partial F_1}{\partial y_2}&\cdots&\dfrac{\partial F_1}{\partial y_m}\\[4pt]\cdots&\cdots&\cdots&\cdots\\[4pt]\dfrac{\partial F_m}{\partial y_1}&\dfrac{\partial F_m}{\partial y_2}&\cdots&\dfrac{\partial F_m}{\partial y_m}\end{bmatrix}\begin{bmatrix}\dfrac{\partial y_1}{\partial x_n}\\[4pt]\dfrac{y_2}{\partial x_n}\\[4pt]\cdots\\[4pt]\dfrac{\partial y_m}{\partial x_n}\end{bmatrix}.$$

Convert the above n matrix equations into a matrix equation ;then by using **Lemma 2,** we obtain

$$
\begin{bmatrix}
\dfrac{\partial f}{\partial x_1} & \dfrac{\partial f}{\partial x_2} & \cdots & \dfrac{\partial f}{\partial x_n} \\
\dfrac{\partial F_1}{\partial x_1} & \dfrac{\partial F_1}{\partial x_2} & \cdots & \dfrac{\partial F_1}{\partial x_n} \\
\cdots & \cdots & \cdots & \cdots \\
\dfrac{\partial F_m}{\partial x_1} & \dfrac{\partial F_m}{\partial x_2} & \cdots & \dfrac{\partial F_m}{\partial x_n}
\end{bmatrix}
= -
\begin{bmatrix}
\dfrac{\partial f}{\partial y_1} & \dfrac{\partial f}{\partial y_2} & \cdots & \dfrac{\partial f}{\partial y_m} \\
\dfrac{\partial F_1}{\partial y_1} & \dfrac{\partial F_1}{\partial y_2} & \cdots & \dfrac{\partial F_1}{\partial y_m} \\
\cdots & \cdots & \cdots & \cdots \\
\dfrac{\partial F_m}{\partial y_1} & \dfrac{\partial F_m}{\partial y_2} & \cdots & \dfrac{\partial F_m}{\partial y_m}
\end{bmatrix}
\begin{bmatrix}
\dfrac{\partial y_1}{\partial x_1} & \dfrac{\partial y_1}{\partial x_2} & \cdots & \dfrac{\partial y_1}{\partial x_n} \\
\dfrac{\partial y_2}{\partial x_1} & \dfrac{\partial y_2}{\partial x_2} & \cdots & \dfrac{\partial y_2}{\partial x_n} \\
\cdots & \cdots & \cdots & \cdots \\
\dfrac{\partial y_m}{\partial x_1} & \dfrac{\partial y_m}{\partial x_2} & \cdots & \dfrac{\partial y_m}{\partial x_n}
\end{bmatrix}
$$

$$
=
\begin{bmatrix}
\dfrac{\partial f}{\partial y_1} & \dfrac{\partial f}{\partial y_2} & \cdots & \dfrac{\partial f}{\partial y_m} \\
\dfrac{\partial F_1}{\partial y_1} & \dfrac{\partial F_1}{\partial y_2} & \cdots & \dfrac{\partial F_1}{\partial y_m} \\
\cdots & \cdots & \cdots & \cdots \\
\dfrac{\partial F_m}{\partial y_1} & \dfrac{\partial F_m}{\partial y_2} & \cdots & \dfrac{\partial F_m}{\partial y_m}
\end{bmatrix}
\begin{bmatrix}
\dfrac{\partial F_1}{\partial y_1} & \dfrac{\partial F_1}{\partial y_2} & \cdots & \dfrac{\partial F_1}{\partial y_m} \\
\dfrac{\partial F_2}{\partial y_1} & \dfrac{\partial F_2}{\partial y_2} & \cdots & \dfrac{\partial F_2}{\partial y_m} \\
\cdots & \cdots & \cdots & \cdots \\
\dfrac{\partial F_m}{\partial y_1} & \dfrac{\partial F_m}{\partial y_2} & \cdots & \dfrac{\partial F_m}{\partial y_m}
\end{bmatrix}^{-1}
\begin{bmatrix}
\dfrac{\partial F_1}{\partial x_1} & \dfrac{\partial F_1}{\partial x_2} & \cdots & \dfrac{\partial F_1}{\partial x_n} \\
\dfrac{\partial F_2}{\partial x_1} & \dfrac{\partial F_2}{\partial x_2} & \cdots & \dfrac{\partial F_2}{\partial x_n} \\
\cdots & \cdots & \cdots & \cdots \\
\dfrac{\partial F_m}{\partial x_1} & \dfrac{\partial F_m}{\partial x_2} & \cdots & \dfrac{\partial F_m}{\partial x_n}
\end{bmatrix}.
$$

The above result of matrix operations is denoted as

$$A_{(m+1)\times n} = B_{(m+1)\times m} \times C^{-1}_{m\times m} \times D_{m\times n},$$

According to the property of matrix multiplication ,

$$r(A_{(m+1)\times n}) \le \min\{r(B_{(m+1)\times m}), r(C_{m\times m}), r(D_{m\times n})\} \le m.$$

$\because \;|C_{m\times m}| \neq 0, r(B_{(m+1)\times m}) = r(C_{m\times m}) = m,$

$\therefore \; r[A_{(m+1)\times n}, B_{(m+1)\times n}] = m,$

Matrix $[A_{(m+1)\times n}, B_{(m+1)\times n}]$ is denoted as $E_{(m+1)\times(m+n)},$

$$\therefore r(E_{(m+1)\times(m+n)}) = r
\begin{bmatrix}
\dfrac{\partial f}{\partial x_1} & \dfrac{\partial f}{\partial x_2} & \cdots & \dfrac{\partial f}{\partial x_n} & \dfrac{\partial f}{\partial y_1} & \cdots & \dfrac{\partial f}{y_m} \\
\dfrac{\partial F_1}{\partial x_1} & \dfrac{\partial F_1}{\partial x_2} & \cdots & \dfrac{\partial F_1}{\partial x_n} & \dfrac{\partial F_1}{\partial y_1} & \cdots & \dfrac{\partial F_1}{\partial y_m} \\
\cdots & \cdots & \cdots & \cdots & \cdots & \cdots & \cdots \\
\dfrac{\partial F_m}{\partial x_1} & \dfrac{\partial F_m}{\partial x_2} & \cdots & \dfrac{\partial F_m}{\partial x_n} & \dfrac{\partial F_m}{\partial y_1} & \cdots & \dfrac{\partial F_m}{\partial y_m}
\end{bmatrix}
= m$$

$\therefore$ the system of m+1 row vectors in the matrix $E_{(m+1)\times(m+n)}$ are linearly dependent;

But the first row vector is non-zero vector ,so there are certain constants $\lambda_1, ..., \lambda_m$,such that

$$N + \lambda_1 N_1 + ... + \lambda_m N_m = \vec{0}.$$

From the above linear combination equation of row vectors ,we can aquire m+n equations of partial derivatives .The m+n equations of partial derivatives are exactly the results ,namely, by using Lagrange multiplier method , we take partial derivatives of the following function

$$L(x_1, ..., x_n, y_1, ..., y_m, \lambda_1, ..., \lambda_m) = f(x_1, ..., x_n, y_1, ..., y_m) + \sum_1^m \lambda_i F_i(x_1, ..., x_n, y_1, ..., y_m)$$

with respect to $x_1, x_2, ..., x_n, y_1, .., y_m$ respectively ,then let m+n partial derivatives be equal to zero.The process of proving Lagrange multiplier method is less rigorous and is actually circular reasoning in the document (WU Chenyu :*A Proof to Lagrange Multiplier* .Journal of Shanxi Datong University .Voi.24.No.3:10) .

(2) Based on the (1) ,because the rank of $E_{(m+1)\times(n+m)}$ is equal to m, the rank of any (m+1)(m+1)subdeterminant in the $E_{(m+1)\times(n+m)}$ is equal to m ,namely ,row vectors in any (m+1)(m+1)subdeterminant are linearly decpendent ,so the value of any (m+1)(m+1) subdeterminant is zero .

Example 20 Solve extreme point ,objective function and constraints are respectively

$$u = x_1^2 + x_2^2 + x_3^2 + x_4^2, x_1 + x_2 + x_3 + x_4 = -1, 2x_1 + x_2 + 3x_3 - 3x_4 = 1$$

Solution:Normal vectors of the surfaces defined by the objective function and constraint equations are as follows

$$N = (x_1, x_2, x_3, x_4) , N_1 = (1,1,1,1), N_2 = (2,1,3,-3) .$$

When the objective function has an extreme value ,the above three vectors are linearly dependent ,there exist real numbers a, b which satisfy

$$N = aN_1 + bN_2,$$

namely

$$(x_1, x_2, x_3, x_4) = (a + 2b, a + b, a + 3b, a - 3b);$$

Pulg them into constraint equations ,obtain $a = -\dfrac{26}{83}, b = \dfrac{7}{83}$,

Then the only extreme point is

$$(x_1, x_2, x_3, x_4) = (-\dfrac{12}{83}, -\dfrac{19}{83}, -\dfrac{5}{83}, -\dfrac{47}{83}).$$

6.Formula of Lagrange multipliers

We do not need Lagrange multiplier method to obtain extreme points of multivariate function under equality constraints now ,but in some cases ,Larange multiplier has a certain meaning ,so it is necessary to calculate multiplier .In the research of general varivational principles in elasticity and plasticity ,Larange multiplier method has been used widely .The scholars from China discussed the problem of whethere Lagrange multiplier was unique or not ,which has been carried out off and for more than 30 years since the dispute between Qian Weichang and Hu Haicang in 1980s .Acoding to the resuits of finite element method ,Liu Shiquan and Liang Lifu expounded that Larange multiplier is unique when there are cloed differential equation grub or algebraic equation grub in some local area (refer to Liu Shiquan and Liang Lifu: *The Application of Larangian Multiplier Mthod in Ganeralized Variational Principles and Finite Element Method*) .But Using outer product of n-1 vetors in n-dimensional space ,we propose that formula of Larane mutilipliers is not unique. In the process of derivation ,the following theorem must be introduced

6.1.Some properties of outer product

In the paper of *Generalization of Outer Product and Application* , some properties of outer product were presented ,now other theorem of outer product are shown as follows :

Theorem 7   Suppose vectoers $\alpha_1, \alpha_2 ..., \alpha_m$ in real m+1 dimensional space are linearly independent .If non-zero real vector α is orthogonal to $\alpha_1, \alpha_2 ..., \alpha_m$,then all vectors which are orthogonal to $\alpha_1, \alpha_2 ..., \alpha_m$ are $k\alpha$,where k is any non –zero real number .

Proof :Denote $\alpha_i = (a_{i1}, a_{i2}, ..., a_{i,m+1})$ ($\alpha = 1, 2, ..., m$), $\alpha = (a_1, a_2, ..., a_{m+1})$. Becuse α is orthogonal to $\alpha_i (i = 1, 2, ..., m)$, $\alpha = (a_1, a_2, ..., a_{m+1})$ is solution vector of the following system of homogeneous linear equations

$$\begin{pmatrix} a_{11} & a_{12} & ... & a_{1,m+1} \\ a_{21} & a_{22} & ... & a_{2,m+1} \\ ... & ... & ... & ... \\ a_{m1} & a_{m2} & ... & a_{m,m+1} \end{pmatrix} \begin{pmatrix} x_1 \\ x_2 \\ ... \\ x_{m+1} \end{pmatrix} = \begin{pmatrix} 0 \\ 0 \\ ... \\ 0 \end{pmatrix} .$$

But the rank of the coefficient matrix is m ,so α is the basic solution of the system of the homogenous linear equation ,thereby all vectors which are orthogonal to $\alpha_1, \alpha_2 ..., \alpha_m$ are $k\alpha$

Theorem 8   Suppose the systems of m+1 dimensioanl real vectors $\alpha_1, \alpha_2 ..., \alpha_m$ and $\beta_1, \beta_2 ..., \beta_m$ both are linearly independent and equivalent to each other ,then the all vectors which are both orthogonal to $\alpha_1, \alpha_2 ..., \alpha_m$ and $\beta_1, \beta_2 ..., \beta_m$ are collinear .

Proof: $\alpha_1, \alpha_2 ..., \alpha_m$ and $\beta_1, \beta_2 ..., \beta_m$ are equivalent to each other ,so there exist real numbers $k_{i,j} (i, j = 1, 2, ..., m)$,so such

$$\alpha_i = k_{i,1}\beta_1 + k_{i,2}\beta_2 + ... + k_{i,m}\beta_m (i = 1, 2, ..., m) .$$

According to Theorem 7, suppose all vectors which are orthogonal to $\beta_1, \beta_2 ..., \beta_m$ are $k\beta$ (where any real number $k \neq 0$) ,then the dop products

$$\alpha_i \cdot k\beta = k(k_{i,1}\beta_1 \cdot \beta + k_{i,2}\beta_2 \cdot \beta + ... + k_{i,m}\beta_m \cdot \beta)$$

$$= k(k_{i,1} \cdot 0 + k_{i,2} \cdot 0 + ... + k_{i,m} \cdot 0) = 0 ,$$

So $k\beta$ are orthogonal to $\alpha_1, \alpha_2 ..., \alpha_m$,thereby all vectors which are orthogonal to $\alpha_1, \alpha_2 ..., \alpha_m$ and $\beta_1, \beta_2 ..., \beta_m$ are collinear .

Corollary   Suppose $\alpha_1, \alpha_2 ..., \alpha_n$ are real vectors in m+1 (n>m)dimensioanl space .If the rank of the vectors is m ,then the all vectors which are orthogonal to $\alpha_1, \alpha_2 ..., \alpha_n$ are collinear .

6.2.Deducing the formula of Lagrange multipliers

New let us deduce the formula of Lagrange multipliers ,which can demonstate that even if at the same extreme point ,the formula of Lagrange multipliers is not unique .

Based on theorem 6.1, $N + \lambda_1 N_1 + ... + \lambda_m N_m = \vec{0}$,

Convert the above equation of vector into the following vector multiplication

$$(1, \lambda_1, \lambda_2, ..., \lambda_m) \begin{bmatrix} N \\ N_1 \\ N_2 \\ ... \\ N_m \end{bmatrix} = (0,0,0,...,0) ,$$

then

$$(1, \lambda_1, \lambda_2, ..., \lambda_m) \begin{bmatrix} \dfrac{\partial f}{\partial x_1} & \dfrac{\partial f}{\partial x_2} & ... & \dfrac{\partial f}{\partial x_n} & \dfrac{\partial f}{\partial y_1} & ... & \dfrac{\partial f}{y_m} \\ \dfrac{\partial F_1}{\partial x_1} & \dfrac{\partial F_1}{\partial x_2} & ... & \dfrac{\partial F_1}{\partial x_n} & \dfrac{\partial F_1}{\partial y_1} & ... & \dfrac{\partial F_1}{\partial y_m} \\ ... & ... & ... & ... & ... & ... & ... \\ \dfrac{\partial F_m}{\partial x_1} & \dfrac{\partial F_m}{\partial x_2} & ... & \dfrac{\partial F_m}{\partial x_n} & \dfrac{\partial F_m}{\partial y_1} & ... & \dfrac{\partial F_m}{\partial y_m} \end{bmatrix} = (0,0,0,...,0) ,$$

The above matrix eqution indicates :at the suspicious point ,the vector $(1, \lambda_1, \lambda_2, ..., \lambda_m)$ is orthogonal to every colum vector in the matrix E .According to theorem 6.1 ,the rank of E is equle to m ,suppose the following subdeterminant

$$\begin{vmatrix} \dfrac{\partial F_1}{\partial y_1} & \dfrac{\partial F_1}{\partial y_2} & ... & \dfrac{\partial F_1}{\partial y_m} \\ \dfrac{\partial F_2}{\partial y_1} & \dfrac{\partial F_2}{\partial y_2} & ... & \dfrac{\partial F_2}{\partial y_m} \\ ... & ... & ... & ... \\ \dfrac{\partial F_m}{\partial y_1} & \dfrac{\partial F_m}{\partial y_2} & ... & \dfrac{\partial F_m}{\partial y_m} \end{vmatrix} \neq 0 ,$$

then in the marix E ,the system of the last m column vectors is maximal linear independent system ,according to lamma 1 and theorem 8 ,the outer product vector of the last m column vectors

$$\begin{vmatrix} i & \dfrac{\partial f}{\partial y_1} & \dfrac{\partial f}{\partial y_2} & ... & \dfrac{\partial f}{\partial y_m} \\ i_1 & \dfrac{\partial F_1}{\partial y_1} & \dfrac{\partial F_1}{\partial y_2} & ... & \dfrac{\partial F_1}{\partial y_m} \\ ... & ... & ... & ... & ... \\ i_m & \dfrac{\partial F_m}{\partial y_1} & \dfrac{\partial F_m}{\partial y_2} & ... & \dfrac{\partial F_m}{\partial y_m} \end{vmatrix} = \begin{bmatrix} B \\ B_1 \\ ... \\ B_m \end{bmatrix}$$

is orthogonal to n+m column cectors in the matrix E (where $B, B_1, ..., B_m$ are subdeterminants are corresponding to $i, i_1, ..., i_m$ respectively) ,so $(B, B_1, ..., B_m)$ and $(1, \lambda_1, ..., \lambda_m)$ are collinear .Because $B \neq 0$, $\dfrac{1}{B} = \dfrac{\lambda_1}{B_1} = ... = \dfrac{\lambda_m}{B_m}$, thereby $\lambda_1 = \dfrac{B_1}{B}, \lambda_2 = \dfrac{B_2}{B}, ..., \lambda_m = \dfrac{B_m}{B}$,these are the Lagrang mulitipliers .

However ,vector maximal independent sub set of column vectors in the marix E is not unique ,outer product of any vector maximal indepentdent sub set of column in the matrix E is collinear to the vector $(1, \lambda_1, ..., \lambda_m)$,so the expressions of Lagrange mulitipliers at a fixed extreme point are not unique ,which ends the dispute between Qian Weichang and Hu Haicang (namely ,if the Lagrange mulitipliers are unique or not) in the 1980s .

Because the vector $(B, B_1, ..., B_m)$ is orthogonal to any column vectors in the matrix E ,

$$B \cdot \frac{\partial f}{\partial x_i} + B_1 \cdot \frac{\partial F_1}{\partial x_i} + B_2 \cdot \frac{\partial F_2}{\partial x_i} + ... + B_m \cdot \frac{\partial F_m}{\partial x_i} = 0 \ (i = 1,2,...,n \ );$$

According to the theorem of Laplace expansion ,the left of every equation is the following m+1 order determinant respectively

$$\begin{vmatrix} \dfrac{\partial f}{\partial x_i} & \dfrac{\partial f}{\partial y_1} & \dfrac{\partial f}{\partial y_2} & \cdots & \dfrac{\partial f}{\partial y_m} \\ \dfrac{\partial F_1}{\partial x_i} & \dfrac{\partial F_1}{\partial y_1} & \dfrac{\partial F_1}{\partial y_2} & \cdots & \dfrac{\partial F_1}{\partial y_m} \\ \cdots & & & \cdots & \\ \dfrac{\partial F_m}{\partial x_i} & \dfrac{\partial F_m}{\partial y_1} & \dfrac{\partial F_m}{\partial y_2} & \cdots & \dfrac{\partial F_m}{\partial y_m} \end{vmatrix} = 0 (i = 1,2,...,n) .$$

Of course ,base on the fact that rank of the matrix E is m ,we can also arrive at a conclusion which the all m+1 order subdeterminants in E are zero ,namely ,any m+1order subdeterminant formed by any m+1 column vector in E is zero .

7.Steps for solving suspicious extreme point of function under equality constraints

According to theorem 6 ,when objective function of n variables under m equality constraints acquires suspicious extreme value , the rank of the matrix $E_{(m+1)\times n}$ formed by the m+1 row vectors is m ,so the all m+1 order subdeterminants formed by any m+1 column vectors in E are zero ,we can obtain n-m subdeterminants of m+1 orders in the $E_{(m+1)\times n}$,then let them be equal to zero ,thereby we obtain n-m equations of n variables.

Note :There are C_n^{m+1} subdeterminants of m+1 orders .

Because elementary transformation for mareix does not chang the rank of matrix ,row elementary transformation for matrix $E_{(m+1)\times n}$ is equivalent to the same row elementary transformation for the all m+1 order subdeterminants .When the susupicious extreme value is acquired , elementary transformation for matrix $E_{(m+1)\times n}$ does not chang the fact that the all m+1 order determinants are zero ,which can simplify expression of $E_{(m+1)\times n}$ and the all expressions of m+1 order subdeterminamts ,This can be illustrated by the folling simple example.

Example 21 Solve susupicious extreme points ,the objective function and equality constrain are as follows respectively

$$u = x^2 + y^2 + z^2, (x - y)^2 - z^2 = 1.$$

Solution: Using the two normal vectors defined by the function and the constraint equation ,we can form 2×3 matrix and operate row elementary transformation to simplify the 2×3 matrix ,namely

$$\begin{bmatrix} 2(x-y) & 2(y-x) & -2z \\ 2x & 2y & 2z \end{bmatrix} \rightarrow \begin{bmatrix} y & x & 2z \\ x & y & z \end{bmatrix} ;$$

Because the objective function obtains suspicious extreme value ,the rank of matrix is one .From the above matrix ,the system of equations

$$\begin{cases} \begin{vmatrix} y & x \\ x & y \end{vmatrix} = 0 \\ \begin{vmatrix} x & 2z \\ y & z \end{vmatrix} = 0 \\ (x-y)^2 - z^2 = 1 \end{cases}$$

is founded ,and is converted into the following two systems of equations

$$\begin{cases} y^2 - x^2 = 0 \\ z = 0 \\ (x-y)^2 - z^2 = 1 \end{cases} , \begin{cases} y^2 - x^2 = 0 \\ x - 2y = 0 \\ (x-y)^2 - z^2 = 1 \end{cases} ,$$

so the suspicious points $P_1(\frac{1}{2}, -\frac{1}{2}, 0)$, $P_2(-\frac{1}{2}, \frac{1}{2}, 0)$ are obtained .

If the third column vector $[2z, z]^T$ is used to form the following system of equations

$$\begin{cases} \begin{vmatrix} y & 2z \\ x & z \end{vmatrix} = 0 \\ \begin{vmatrix} x & 2z \\ y & z \end{vmatrix} = 0 \\ (x-y)^2 - z^2 = 1 \end{cases} ,$$

we can only get the value of z ,but can not get the values of x, y .Because vector $[2z, z]^T = [0,0]^T$ at the

point P_1 or the point P_2 is a zero vector ,and a zero vector is linearly dependent to itself ,the inner products

between the vector $[2z, z]^T = [0,0]^T$ and the other two column vectors are both zero .There does not exist

implicit function $z = z(x, y)$ at P_1, P_2 .

The process of solving Example 21 manifests :(1)elememtary transformation for matrix can simplify calculation of solving exemem points of function under equality constraints can simplify calculation ,(2)choice of m column vectors in the matrix $E_{(m+1)\times n}$ is also important ,otherwise we can only get the value of a certain unknown ,can not get the values of all the unknowns .But in this case ,the problem to solving extreme values of a function of n variables under equality constraints can be changed into the problem to solving extreme values of function of n-1 variables .

So ,the steps of solving suspicious extreme points of function of n variables under m equality constraints are as follows:

(1)Use normal vectors of the surfaces (or curves) defined by the objective function and constraint equations to form a $E_{(m+1)\times n}$;

(2)Operate elementary transformation for the matrix $E_{(m+1)\times n}$ to simplify the matrix (if necessary);

(3)Selecting a column vector maximal independent sub set of of the matrix $E_{(m+1)\times n}$,use it and every one of the other column vectors to form m-n subdeterminants of m+1 order respectively ,let them be zero ,and obtain n-m equations ;

(4)Solve system of equations (n×n) to acquire suspicious points.

Example 22: Solve suspicious points ,objective function and constraint equations are as follows

$$u = x^3 + y^3 + z^3 - 2xyz , x^2 + y^2 + z^2 = 1.$$

Solution :Use two normal vectors to form a matrix

$$\begin{pmatrix} 3x^2 - 2yz & 3y^2 - 2xz & 3z^2 - 2xy \\ x & y & z \end{pmatrix},$$

when the objective function obtains suspicious extreme value ,the rank of the above matrix is one ,and then a system of equations

$$\begin{cases} (3x^2 - 2yz)y - (3y^2 - 2xz)x = 0 \\ (3x^2 - 2yz)z - (3z^2 - 2xy)x = 0 \end{cases},$$

is obtain ,namely

$$\begin{cases} (x - y)(3xy + 2xz + 2yz) = 0 \\ (x - z)(3xz + 2xy + 2yz) = 0 \end{cases},$$

It is changed into the following four independent systems of equations

$$\begin{cases} x - y = 0 \\ x - z = 0 \\ x^2 + y^2 + z^2 = 1 \end{cases}, \begin{cases} x - y = 0 \\ 3xz + 2xy + 2yz = 0 \\ x^2 + y^2 + z^2 = 1 \end{cases}, \begin{cases} 3xy + 2xz + 2yz = 0 \\ x - z = 0 \\ x^2 + y^2 + z^2 = 1 \end{cases}, \begin{cases} 3xy + 2xz + 2yz = 0 \\ 3xz + 2xy + 2yz = 0 \\ x^2 + y^2 + z^2 = 1 \end{cases}.$$

Solving above systems of equations ,thereby the suspicious points are as follows .

$$\pm(\frac{\sqrt{3}}{3}, \frac{\sqrt{3}}{3}, \frac{\sqrt{3}}{3}) , (0,0,\pm1) , \pm(\frac{5}{3\sqrt{6}}, \frac{5}{3\sqrt{6}}, -\frac{2}{3\sqrt{6}}) , (0,\pm1,0) , \pm(\frac{5}{3\sqrt{6}}, -\frac{2}{3\sqrt{6}}, \frac{5}{3\sqrt{6}}) ,$$

$$(\pm1,0,0) , \pm(-\frac{2}{3\sqrt{6}}, \frac{5}{3\sqrt{6}}, \frac{5}{3\sqrt{6}})$$

8.Method of drop element for multivariate function under equality constraints

8.1.Perfection for implicit function theorem

From the process of solving example 21 ,lemma 2 ,namely implicit function theorem has imperfection ,the last m variables are not always functions of the other n-m variables .In fact ,when the m equations have infinitely many common solutions and Jacobian matrix is row full rank at some point of solution (namely ,there is a nonzero subdeterminaint of m orders) ,m variables of the n variables are implicit functios of the other n-m variables in neighbourhood of the point P .Thereby the implicit function theorem in lamma is improved as

Theorem 11   Given the system of equations

$$F_i(x_1,x_2,...,x_n) = 0, i = 1,2,...,m(n > m).$$

Suppose at some point $P(x_1^0, x_2^0,...,x_n^0)$,the following three conditions are simultaneously satisfied

(1) $F_i(x_1^0, x_2^0,...,x_n^0) = 0, i = 1,2,...,m$,

(2)Every of the functions $F_i, i = 1,2,...,m$ has comtinous first-order partial derivatives in the neighbourhood of the point P ,

(3)there exist m linearly independent column vectors in the following Jacobian matrix

$$\frac{\partial(F_1,F_2,...,F_m)}{\partial(x_1,x_2,...,x_n)} = \begin{bmatrix} \dfrac{\partial F_1}{\partial x_1} & \dfrac{\partial F_1}{\partial x_2} & \cdots & \dfrac{\partial F_1}{\partial x_n} \\ \dfrac{\partial F_2}{\partial x_1} & \dfrac{\partial F_2}{\partial x_2} & \cdots & \dfrac{\partial F_2}{\partial x_n} \\ \cdots & \cdots & \cdots & \cdots \\ \dfrac{\partial F_m}{\partial x_1} & \dfrac{\partial F_m}{\partial x_2} & \cdots & \dfrac{\partial F_m}{\partial x_n} \end{bmatrix};$$

Then the m variables corresponding to taking partial derivatives   are implicit functions of the other n-m variables.

Suppose the first m column vectors are linearly dependent ,namely the following determinant

$$\begin{vmatrix} \dfrac{\partial F_1}{\partial x_1} & \dfrac{\partial F_1}{\partial x_2} & \cdots & \dfrac{\partial F_1}{\partial x_m} \\ \dfrac{\partial F_2}{\partial x_1} & \dfrac{\partial F_2}{\partial x_2} & \cdots & \dfrac{\partial F_2}{\partial x_m} \\ \cdots & \cdots & \cdots & \cdots \\ \dfrac{\partial F_m}{\partial x_1} & \dfrac{\partial F_m}{\partial x_2} & \cdots & \dfrac{\partial F_m}{\partial x_m} \end{vmatrix} \neq 0,$$

then:

(1)the syste of implicit functions

$$x_i = x_i(x_{m+1}, x_{m+2},...,x_n), i = 1,2,...,m$$

are only defined by the system of equations (also called to be vector-valued functions) in the neighbourhood of the point P ,the system of implicit functions also satisfy the following system of equations

$$F_i[x_1(x_{m+1},x_{m+2},...,x_n),x_2(x_{m+1},x_{m+2},...,x_n),\ ...,x_m(x_{m+1},x_{m+2},...,x_n),x_{m+1},x_{m+2},...,x_n]=0,$$

and $x_i^0 = x_i(x_{m+1}^0,x_{m+2}^0,...,x_n^0), i=1,2,...,m$;

(2)the system of implicit functions have contiuous first order partial derivatives in the neighbourhood of the point $(x_{m+1}^0,x_{m+2}^0,...,x_n^0)$.

Of course ,if the last m column vectors $x_{n-m+1},x_{n-m+2},...,x_n$ are linearly independent ,the m variables

$x_{n-m+1},x_{n-m+2},...,x_n$ are implicit functions of the first variables $x_1,x_2,...,x_{n-m}$ .In a word ,once there exist m linearly independent column vectors with respect partial derivatives in the matrix ,the m variables are implicit functions of the other n-m variables.

The improved implicit function theorem is concise and is used easily(the theorem is further improved in later).

8.2.Approach to dimension reducing for multivariate function under equality constraints (or called method of descending dimension)

In the process of solving linear equations ,substitution method is used to reduce unknowns .In the same way ,substitution method can be also used to some questions of extreme value of multivariate functions under equality constraints .Suppose ,for example ,the objective function and constraints are as follows respectively

$$u=f(x_1,x_2,...,x_n,y_1,y_2,...,y_m),$$

$$\begin{cases} a_{11}x_1+...+a_{1n}x_n+b_{11}y_1+...+b_{1m}y_m=b_1 \\ a_{21}x_1+...+a_{2n}x_n+b_{21}y_1+...+b_{2m}y_m=b_2 \\ ... \\ a_{m1}x_1+...+a_{mn}x_n+b_{m1}y_1+...+b_{mm}y_m=b_m \end{cases} ,\text{where} \begin{vmatrix} b_{11} & b_{12} & ... & b_{1m} \\ b_{21} & b_{22} & ... & b_{2m} \\ ... & ... & ... & ... \\ b_{m1} & b_{m2} & ... & b_{mm} \end{vmatrix} \neq 0 .$$

then according to implicit function theorem ,the system of equations defines the functions $y_1,y_2,...,y_m$ of the

variables $x_1,x_2,...,x_n$ .Using elimination by addition or subtraction ,the expressions of the $y_1,y_2,...,y_m$ in

terms of $x_1,x_2,...,x_n$ are obtained ,then the function of m+n variables under equality constraints is changed into the function of n variables without any constraints.

But for the example 21 ,in the process of solving extreme points ,expression $z^2=(x-y)^2-1$ can not be

substituted into the objective function $u=x^2+y^2+z^2$,this is because the constraint equation

$z^2=(x-y)^2-1$ defines $z=\pm\sqrt{(x-y)^2-1}$, wich is not the only expression of z about variables x,y .

So ,with the combination of example ,the sufficient condition can be induced as follows :

If from the system of m constraint equations , p variables can be expressed as the only explicit functions of the m-p other variables , the original objective function of n variables under m equality constraints can be conveted into the function of n-p variables under m-p equality constraints .

8.3. Converting Multivariate function under equality constraints into function without any constraints

In the theory of system of linear equations of n varivaribles ,suppose the rank of the coefficient matrix and the rank of the augmented matrix are both m (m<n) ,then the general solution is linear expression of the remaining n-m variables .In this case ,it is considered that the system of equations of n variables defines the expressins of n explicit functions .In fact ,general equation of circle is converted into system of equations with a parameter , general equation of surface is changed into system of equations with two parameters ,and so on ,the all similar cases can indicate that some equations of n variables or some systems of equations of n variables can define expressions of n explicit functions .So implicit function theorem is developed as the following explicit function theorem :

Proposition 1 (explicit function proposition) Suppose system of equations of n variables

$$F_i(x_1, x_2, ..., x_n) = 0 (i = 1, 2, ..., m, m < n)$$

satisfy simultaneously the following three conditions:

(1)has infinitely many solutions $(x_1, x_2, ..., x_n)$ (the solution set is denoted as Ω);

(2)every of the function $F_i(x_1, x_2, ..., x_n), i = 1, 2, ..., m$ has continuous first partial derivatives;

(3) $\forall (x_1, x_2, ..., x_n) \in \Omega$,there exist m linealy independent column vectors in Jacobian

$\dfrac{\partial(F_1, F_2, ..., F_m)}{\partial(x_1, x_2, ..., x_n)}$ (suppose the first m column vectors are linearly independent).

Then the system defines the only system of explict functions (or called to be vector-valued function)

$$\begin{cases} x_1 = x_1(t_1, t_2, ..., t_{n-m}) \\ x_2 = x_2(t_1, t_2, ..., t_{n-m}) \\ ...... \\ x_n = x_n(t_1, t_2, ..., t_{n-m}) \end{cases}, (t_1, t_2, ..., t_{n-m}) \in \Omega' \subset R^{n-m},$$

and the system of functions (vector-valued function) satisfy the following every equaton

$$F_i(x_1(t_1, t_2, ..., t_{n-m}), x_2(t_1, t_2, ..., t_{n-m}), ..., x_n(t_1, t_2, ..., t_{n-m})) = 0, i = 1, 2, ..., m,$$

and the system of functions has continuous first (partial) derivatives in the Ω' .

Note :Proposition 1 has very broadly applications in other branches of mathematics .

Example 23 System $\begin{cases} x^2 + y^2 + z^2 = R^2 \\ x + y + z = 0 \end{cases}$ defines the only three explicit functions

$$x = \frac{R}{\sqrt{6}} \sin t + \frac{R}{\sqrt{2}} \cos t, \ y = \frac{R}{\sqrt{6}} \sin t - \frac{R}{\sqrt{2}} \cos t, z = -\frac{2R}{\sqrt{6}} \sin t$$

on the interval $[0, 2\pi]$,they have firs-ordert continuous derivatives

Example 24 Suppose $x, y, z \geq 1$, then the following system of equations

$$\begin{cases} 11x^2 - 9y^2 = 2 \\ 40y^2 - 11z^2 = 29 \end{cases}$$

defines the following three explicit functions

$$x = \sqrt{36t+1}, \ y = \sqrt{44t+1}, \ z = \sqrt{160t+1} \ (t \geq 0).$$

So ,based on proposition 1 ,the following conclusion can be derived

Proposition 2 Given function of n variables under m equality constraints (m<n) .If the constraint system of equatons satisfy the above the condition of proposition 1 ,the objective function can theoretically be converted into function of n-m variables without any constraint.

But at present condition of mathematical knowledge and mathematical tool ,it is hard enough to obtain expressions of explicit functions from nonlinear equaton of n variables or from syste of nonlinear equations (with few exceptions) .Even if expressions of explicit functions can be obtained from nonlinear equaton of n variables or from syste of nonlinear equations ,and the objective function is converted into function without any constraints ,it is also difficult to get extreme point ,even it is more diffcult .But the two propositions have theoretical significance ,which probably offer help to deal with other issues .

9.Discriminations for suspicious extreme point

9.1.Discriminattions for ordinary extreme point

9.1.1.The other two discrimination method for extreme point of unary function

(1)if function $f(x)$ has second derivative at the point x_0 ,and $f'(x_0) = 0$, $f''(x_0) \neq 0$; then x_0 is extreme point .In this case ,if $f''(x_0) > 0$,then x_0 is minimum point ;if $f''(x_0) < 0$,then x_0 is maximum point .

(2)if function $f(x)$ has n th-order derivatives at the point x_0 $(n \geq 3)$, and

$$f'(x_0) = f''(x_0) = ... = f^{(n-1)}(x_0), \text{but } f^{(n)}(x_0) \neq 0 .\text{In this case ,if } n \text{ is an odd number ,then} x_0 \text{ is not}$$

point ;if n is even number ,and $f^{(n)}(x_0) > 0$, x_0 is minimum point ;if n is even and $f^{(n)}(x_0) < 0$, x_0 is maximum point.

9.1.2 Discrimination method for binary function

Suppose $f(x, y)$ have second continuous derivatives at point (x_0, y_0) ,denote $A = f_{xx}(x_0, y_0)$,

$B = f_{xy}(x_0, y_0)$, $C = f_{yy}(x_0, y_0)$, $H = \begin{vmatrix} A & B \\ B & C \end{vmatrix} = AC - B^2$,then

1)If $H > 0, A > 0$, (x_0, y_0) is minimum point;

2)If $H > 0, A < 0, (x_0, y_0)$ maximuu point;

3)If $H < 0, (x_0, y_0)$ is not extreme point;

4)If $H = 0$,can not determine whether (x_0, y_0) is point or not .

9.1.3.Discrimination method for function of n variables without constraints

Suppose function $u = f(x_1, x_2, ..., x_n)$ has continuous second-order partial deriavateves at the suspicious extreme point $P_0(x_1^0, x_2^0, ..., x_n^0)$ .If the Hessian matrix

$$H = \left(\frac{\partial^2 f}{\partial x_i \partial x_j} \right) |_{P_0},$$

is positive definite ,P_0 is local minimum point ;if H is negative definite, P_0 is local maximum point ;if H has both positive and negative eigenvalue, then P_0 is not extreme point ;if H is inconclusive , can not determine whether P_0 is point or not.

9.2.Second derivative method for multivariate function under equality constraints

In current textbooks and reference books concerning differential calculus of multivariate function ,there are no efficient methods to discriminate suspious extreme point of multivariate function under equality constraints .

Undoubtedly ,geometric interpretations of multivariate function and multivariate equation can be the best tool to discriminate suspiciuous points ,but if the geometric shape of surface (curve) defined by a multivariate function or a multivariate equation represents a curve or a surface can not be determined ,we have to find other discriminated method .Acorrding to proposition 2 , function of n variables under m equality constraints can theoretically be converted into function of n-m variables without any constraint .Consequently ,implicit function theorem can be used to discriminate suspicious point of function under equality constraints .

Suppose point $P_0(x_1^0, x_2^0, ..., x_n^0, y_1^0, y_2^0, ..., y_m^0)$ is suspicious point of the objective function

$$u = f(x_1, x_2, ..., x_n, y_1, y_2, ..., y_m)$$

under the equality constraints

$$F_i(x_1, x_2, ..., x_n, y_1, y_2, ..., y_m) = 0, i = 1, 2, ..., m.$$

If at the point P_0 , the following Jacobian determinant

$$\left|\frac{\partial(F_1,F_2,...,F_m)}{\partial(y_1,y_2,...,y_m)}\right|_P \neq 0;$$

Then according to implicit function theorem and the obove proposition 2, in the neighbourhood of the point $P'(x_1^0,x_2^0,...,x_n^0)$, the original objective function can be converted into the following function of n variables without constraints

$$w(x_1,...,x_n) = f[x_1,x_2,...,x_n,y_1(x_1,x_2,...,x_n),y_2(x_1,x_2,...,x_n),...,y_m(x_1,x_2,...,x_n)].$$

where $y_1,y_2,...,y_m$ are intermediate variables . From the constraint system of equations

$$F_i(x_1,x_2,...,x_n,y_1,y_2,...,y_m) = 0, i = 1,2,...,m,$$

We obtain partial derivatives

$$\frac{\partial y_1}{\partial x_j},\frac{\partial y_2}{\partial x_j},...,\frac{y_m}{\partial x_j} = 0, j = 1,2,...,n \quad,$$

Then we take first-order partial derivatives of function of n variables $w(x_1,x_2,...,x_n)$

$$\frac{\partial w}{\partial x_j} = \frac{\partial f}{\partial x_j} + \frac{\partial f}{\partial y_j}\frac{\partial y_j}{\partial x_j}, j = 1,2,...,n \quad,$$

then we take second-order partial derivatives of function of n variables $w(x_1,x_2,...,x_n)$

$$\frac{\partial}{\partial x_1}(\frac{\partial w}{\partial x_j}),\frac{\partial}{\partial x_2}(\frac{\partial w}{\partial x_j}),...,\frac{\partial}{\partial x_n}(\frac{\partial w}{\partial x_j}), j = 1,2,...,n \quad.$$

Then we can eatablish the following Hessian matrix

$$H = \begin{pmatrix} \dfrac{\partial^2 w}{\partial x_1^2} & \dfrac{\partial^2 w}{\partial x_1 \partial x_2} & \cdots & \dfrac{\partial^2 w}{\partial x_1 \partial x_n} \\ \dfrac{\partial^2 w}{\partial x_2 \partial x_1} & \dfrac{\partial^2 w}{\partial x_2^2} & \cdots & \dfrac{\partial^2 w}{\partial x_2 \partial x_n} \\ \cdots & \cdots & \cdots & \cdots \\ \dfrac{\partial^2 w}{\partial x_n \partial x_1} & \dfrac{\partial^2 w}{\partial x_n \partial x_2} & \cdots & \dfrac{\partial^2 w}{\partial x_n^2} \end{pmatrix}$$

According to crimination method for function of n variables without any constraints ,we can speculare if the point $P_0(x_1^0,x_2^0,...,x_n^0,y_1^0,y_2^0,..,y_m^0)$ is an extreme point or not .

Of course ,if m=n-1 ,then the Hessian matrix is a number .

Example 25: Discriminate $P_1(\frac{1}{2},-\frac{1}{2},0)$ and $P_2(-\frac{1}{2},\frac{1}{2},0)$ .Function $u = x^2 + y^2 + z^2$ under the equality constraint $(x-y)^2 - z^2 = 1$.

Solution : Let $F(x,y,z) = (x-y)^2 - z^2 - 1$.

Because $F_z = -2z = 0, F_x = 2(x - y) = 1 \neq 0$ at the point $P_1(\frac{1}{2}, -\frac{1}{2}, 0)$, there exists implicit function

$x = x(y, z)$ in the neighbourhood of P_1 ,and the original objective function can be converted into the following function without constraint

$$f = u[x(y, z), y, z] = x^2 + y^2 + z^2 \ ,$$

on the neighbourhood of the point $P_1'(y_0, z_0) = (-\frac{1}{2}, 0)$,where x is an intermediate variable .

From the constraint equation ,obtain partial deriavatives

$$\frac{\partial x}{\partial y} = -\frac{F_y}{F_x} = 1, \frac{\partial x}{\partial z} = -\frac{F_z}{F_x} = \frac{z}{x - y};$$

then

$$\frac{\partial f}{\partial y} = 2y + \frac{\partial u}{\partial x}\frac{\partial x}{\partial y} = 2x + 2y, \frac{\partial f}{\partial z} = 2z + \frac{\partial u}{\partial x}\frac{\partial x}{\partial z} = 2x\frac{z}{x - y} + 2z = 4z + \frac{2yz}{x - y};$$

And then second-order partial deriavatives are as follows

$$\frac{\partial^2 f}{\partial y \partial y} = \frac{\partial(2x + 2y)}{\partial y} = 2\frac{\partial x}{\partial y} + 2 = 4,$$

$$\frac{\partial^2 f}{\partial y \partial z} = \frac{\partial(2x + 2y)}{\partial z} = 2\frac{\partial x}{\partial z} = \frac{2z}{x - y},$$

$$\frac{\partial^2 f}{\partial z \partial y} = \frac{\partial}{\partial y}(4z + \frac{2yz}{x - y}) = 0 + \frac{2z}{x - y} + 2yz\frac{\partial}{\partial y}(\frac{1}{x - y}),$$

$$\frac{\partial^2 f}{\partial z \partial z} = \frac{\partial}{\partial z}(4z + \frac{2yz}{x - y}) = 4 + \frac{2y}{x - y} + 2yz\frac{\partial}{\partial z}(\frac{1}{x - y});$$

At the P_1

$$\frac{\partial^2 f}{\partial y \partial y} = 4, \frac{\partial^2 f}{\partial y \partial z} = 0, \frac{\partial^2 f}{\partial z \partial y} = 0, \frac{\partial^2 f}{\partial z \partial z} = \frac{7}{2};$$

The Hessian matrix

$$\begin{pmatrix} \dfrac{\partial^2 f}{\partial y \partial y} & \dfrac{\partial^2 f}{\partial y \partial z} \\ \dfrac{\partial^2 f}{\partial z \partial y} & \dfrac{\partial^2 f}{\partial z \partial z} \end{pmatrix} = \begin{pmatrix} 4 & 0 \\ 0 & \dfrac{7}{2} \end{pmatrix}$$

is positive definite at P_1 .In a similar way ,the Hessian is positive definite at P_2 ,so P_1 and P_2 both are mimimum points .

On the side ,because the objective function denots a moving sphere and the constraint eqution denots a static oblique cylinder ,while the moving sphere is increasing from a point ,the position relationship between the sphere and the cylinder , in order ,is apart away ,tangent ,intersectant .So the tangent point (namely extreme point) is minimum point .

Example 26: Solve and discriminate the exremum of Function $f(x, y) = x^2 y$ under the equality constraint $x^2 + y^2 = 1$.

Solution :According to theorem 4 ,when the objective function obtains extreme value , vectors $N = (2xy, x^2), N_0 = (x, y)$ are collinear , from the following system of equations

$$\begin{cases} \begin{vmatrix} 2xy & x^2 \\ x & y \end{vmatrix} = 0 \\ x^2 + y^2 = 1 \end{cases}$$

Obtain six suspicious points $(0, \pm 1), (\pm\sqrt{\frac{2}{3}}, \pm\sqrt{\frac{1}{3}})$.

Let $F = x^2 + y^2 - 1$,because $F_x = 2xy, F_y = x^2 \neq 0$ at those suspicious points ,the constraint equation difines implicit functions $y = y(x)$ in those neighbourhoods of those points ,so original objective function $f(x, y) = x^2 y$ is converted into the following function without any constraint conditions

$$u(x) = f(x, y(x)) = x^2 y \quad , \quad y = y(x)$$

where y is an intermediate variable ,from constraint equation ,obtain

$$y_x = -\frac{F_x}{F_y} = -\frac{x}{y} \quad ;$$

Then the first order deriavative of function $u(x)$ is

$$\frac{du}{dx} = 2xy + x^2 y_x = 2xy - \frac{x^3}{y} \quad ,$$

And then the second order deriavative of function $u(x)$ is

$$\frac{d^2u}{dx^2} = 2y + 2xy_x - \frac{3x^2}{y} - x^3(-\frac{1}{y^2})y_x = 2y - \frac{5x^2}{y} - \frac{x^4}{y^3} \quad ;$$

At the point $(0, 1)$, $\frac{d^2u}{dx^2} = 2 > 0$,so the point is local minimum point ,

at the point $(0, -1)$, $\frac{d^2u}{dx^2} = -2 < 0$,so the point is local maxmum point , at the points $(\pm\sqrt{\dfrac{2}{3}},\sqrt{\dfrac{1}{3}})$, $\dfrac{d^2u}{dx^2}=-\dfrac{12}{\sqrt{3}}<0$,so the two points are local maxmum points ,

at the points $(\pm\sqrt{\dfrac{2}{3}},-\sqrt{\dfrac{1}{3}})$, $\dfrac{d^2u}{dx^2}=\dfrac{12}{\sqrt{3}}>0$,so the two points are local minimum points .

Thereby the objective obtains minimum $-\dfrac{2\sqrt{3}}{9}$ at points $(\pm\sqrt{\dfrac{2}{3}},-\sqrt{\dfrac{1}{3}})$,and obtains maxmum $\dfrac{2\sqrt{3}}{9}$ at

points $(\pm\sqrt{\dfrac{2}{3}},\sqrt{\dfrac{1}{3}})$.

On the side , objective function $f(x,y)=x^2y$ expresses a moving curve with two branches. When $t>0$

the moving curve $x^2y=t$ is above the x-axis ,which asymptotic lines is x-axis and positive y-axis ;when $t<0$

this moving curve is below the x-axis , which asymptotic lines is x-axis and negative y-axis .

$$\lim_{t\to 0^+}\{l:x^2y=t,t>0\}=\{l_1:\text{x-axis and positive y-axis}\},$$

$$\lim_{t\to 0^-}\{l:x^2y=t,t<0\}=\{l_2:\text{x-axis and negative y-axis}\}.$$

When $t=\dfrac{2\sqrt{3}}{9}$ the moving curve is tangent to the static curve $x^2+y^2=1$,when $0<t<\dfrac{2\sqrt{3}}{9}$ this

moving and this static curve are intersectant ;when $t=-\dfrac{2\sqrt{3}}{9}$ the moving curve is also tangent to the static

curve , when $-\dfrac{2\sqrt{3}}{9}<t<0$ this moving curve and this static curve are also intersectant .

Let l_0 denotes curve defined by eqation $x^2+y^2=1$,then

$$\lim_{t\to 0^+}l\cap l_0=\{(0,1),(\pm1,0)\}\ ,\lim_{t\to 0^-}l\cap l_0=\{(0,-1),(\pm1,0)\}\ .$$

Example 27 :Solve and discriminate the exremum of Function $f(x,y,z)=x^4+y^4+z^4$ under the

equality constraint $xyz=1$.

Solution :According to theorem 2 ,when objective function obtains extreme value , vectors

$N=(x^3,y^3,z^3),N_0=(yz,xz,xy)$ are collinear .Namely

$$\dfrac{x^3}{yz}=\dfrac{y^3}{xz}=\dfrac{z^3}{xy}\ ,$$

$$\frac{x^4}{xyz} = \frac{y^4}{xyz} = \frac{z^4}{xyz} \quad,$$

$$|x| = |y| = |z| \quad;$$

Solve the following system of equations

$$\begin{cases} |x| = |y| = |z| \\ xyz = 1 \end{cases}$$

Obtain four suspicious points $(1,1,1), (1,-1,-1), (-1,-1,1), (-1,1,-1)$;

at suspicious extreme points ,then equation $xyz = 1$ defines implicit function $z = z(x,y)$ in neibourhoods of suspicious points ,thereby objective function

$$f(x, y, z) = x^4 + y^4 + z^4$$

is concerted into the following function without any constraints

$$w(x, y) = x^4 + y^4 + z^4 \quad, \quad z = z(x, y)$$

where z is an intermediate variable ,from constraint equation ,obtain

$$z_x = -\frac{F_x}{F_z} = -\frac{z}{x} \quad, \quad z_y = -\frac{F_y}{F_z} = -\frac{z}{y} \quad;$$

The partial deriavatives of function $w(x, y)$ are

$$w_x = 4x^3 + 4z^3 z_x = 4x^3 - \frac{4z^4}{x} \quad, \quad w_y = 4y^3 + 4z^3 z_y = 4y^3 - \frac{4z^4}{y} \quad;$$

The second-order partial deriavatives of $w(x, y)$ are as follows

$$w_{xx} = 12x^2 + \frac{20z^4}{x^2} \quad, \quad w_{xy} = \frac{16z^4}{x^2} \quad, \quad w_{yx} = \frac{16z^4}{y^2} \quad, \quad w_{yy} = 12y^2 + \frac{20z^4}{y^2} \quad;$$

The Hessian matrix at point $(1,1,1)$

$$\begin{pmatrix} w_{xx} & w_{xy} \\ w_{yx} & w_{yy} \end{pmatrix} = \begin{pmatrix} 32 & 16 \\ 16 & 32 \end{pmatrix} \quad,$$

is positive definite .The other three Hessian matrixs are also positive definite ,so the four points are all mimimum points .Thereby $f_{\min} = 3$.

But the above process of calculating first-order partial derivatives and second partial derivatives is very complicated .It is more complicated to calculate first-order partial derivatives from constraint system of equations .

The process of answering example 25 and example 26 show that if we know the shape of a curve (or surface) described by a multivariate equation and the variation of a moving curve (or surface) described by a multivariate function ,we can very easely discriminate extreme points .

10.Distance among point ,curves , surfaces

Calculation of the distance between point and straight line ,between straight line and circle in the plane ,or between point and plane is quite familiar to us .Now ,calculation of distance will be generalized to general cases ,namely ,calculation of distance between point and curve ,between point and surface ,between curves ,between surfaces ,between curve and surface will be discussed .

10.1.Distance among point and curve in the plane

Definition 10: Fixed point $P(a,b) \notin l$, $Q_0(x_0, y_0) \in l$ .if $\forall Q(x, y) \in l$, inequation $|PQ| \geq |PQ_0|$

always stands ,then the length of line segment PQ_0 is called to be distance from $P(a,b)$ to   curve

l .Namely ,the distance from a point P to curve l is the length of the shortest line segent among the all line segments that connect P and l .

Theorem 12: Suppose $l : f(x, y) = 0$ is a smooth curve ,fixed

$P(x_0, y_0) \notin l, Q(x, y) \in l$.

If $|PQ| = \sqrt{(x - x_0)^2 + (y - y_0)^2}$

obtains mimimum value ,then vector $\overrightarrow{PQ} = (x - x_0, y - y_0)$ and the normal vector (f_x, f_y) are collinear.

Proof :Calculating the distance from point P to curve $l : f(x, y) = 0$ is solving minimum value of the

following function

$$z(x, y) = \sqrt{(x - x_0)^2 + (y - y_0)^2}$$

under the equality constraint $f(x, y) = 0$.

Because the minimum point of function $u(x, y) = (x - x_0)^2 + (y - y_0)^2$ is the minimum point of the

$z(x, y)$ .Acooding to theorem 1 ,when function $u(x, y)$ obtains exreme value ,the vector

$N = (u_x, u_y) = 2(x - x_0, y - y_0)$ and the vector $N = (f_x, f_y)$ are collinear .

Note:If equation of the curve l can be converted into system of equations with a parameter

$x = x(t), y = y(t)$, then the question is changed into the question ,namely solving minimum value of the

function without constraint

$$u(t) = [x(t) - x_0]^2 + [y(t) - y_0]^2 .$$

Definition 9: $P_0 \in l_1, Q_0 \in l_2$,if $\forall P(x, y) \in l_1, \forall Q(x, y) \in l_2$,inequation $|PQ| \geq |P_0Q_0|$ always

stands ,then the length of sement line P_0Q_0 is called the distance between l_1, l_2 .Namely ,distance is the length of the shortest line segment among the all line segments that connect curves l_1, l_2 .

Note:Similarly ,we can define the distance between two surface ,and distance between surface and curve .

Theorem13: Suppose smooth curves $l_1 : f(x, y) = 0; l_2 : g(x, y) = 0$ are disjoint ,points

$P(x_1, y_1) \in l_1, Q(x_2, y_2) \in l_2$ .When the length of line segment PQ is the shortest, then the

vectors $\overrightarrow{PQ} = (x_1 - x_2, y_1 - y_2)$, $N_1 = (f_{x_1}, f_{y_1})$, $N_2 = (g_{x_2}, g_{y_2})$ are collinear.

Proof :When the function of four variables under equality constraints

$$u(x_1, x_2, y_1, y_2) = (x_1 - x_2)^2 + (y_1 - y_2)^2 , f(x_1, y_1) = 0, g(x_2, y_2) = 0$$

obtains minimum value , PQ is the shortest .

According to theorem 6 ,when $u(x_1, x_2, y_1, y_2)$ obtains extreme value ,the three vectors

$$N_1' = (f_{x_1}, f_{y_1}, 0, 0), N_2' = (0, 0, g_{x_2}, g_{y_2}), N' = (x_1 - x_2, y_1 - y_2, x_2 - x_1, y_2 - y_1)$$

are linearly dependent ,namely rank of the matrix

$$\begin{pmatrix} f_{x_1} & f_{y_1} & 0 & 0 \\ 0 & 0 & g_{x_2} & g_{y_2} \\ x_1 - x_2 & y_1 - y_2 & x_2 - x_1 & y_2 - y_1 \end{pmatrix}$$

is two, so two subdeterminants

$$\begin{vmatrix} f_{x_1} & f_{y_1} & 0 \\ 0 & 0 & g_{x_2} \\ x_1 - x_2 & y_1 - y_2 & x_2 - x_1 \end{vmatrix} = 0, \begin{vmatrix} f_{y_1} & 0 & 0 \\ 0 & g_{x_2} & g_{y_2} \\ y_1 - y_2 & x_2 - x_1 & y_2 - y_1 \end{vmatrix} = 0,$$

Namely

$$g_{x_2} \begin{vmatrix} f_{x_1} & f_{y_1} \\ x_1 - x_2 & y_1 - y_2 \end{vmatrix} = 0, f_{y_1} \begin{vmatrix} g_{x_2} & g_{y_2} \\ x_2 - x_1 & y_2 - y_1 \end{vmatrix} = 0,$$

Thereby $(f_{x_1}, f_{y_1}), (x_1 - x_2, y_1 - y_2), (g_{x_2}, g_{y_2})$ are colinear.

Theorem13 indicates that when the line segment which connects two curves is the shortest ,the vector of connecting line , normal vectors at two connecting points are collinear .Thereby this theorem provides a method of calculating the distance between two curves ,namely by solving the following system of equations of four variables

$$\begin{cases} f(x_1, y_1) = 0 \\ g(x_2, y_2) = 0 \\ f_{x_1} \cdot (y_1 - y_2) = f_{y_1} \cdot (x_1 - x_2) \\ g_{x_2} \cdot (y_2 - y_1) = g_{y_2} \cdot (x_2 - x_1) \end{cases},$$

coordinates of the two connecting points which is on the two curves respectively can be obtained.

If the equation of a curve can be converted into system of equations with a parameter ,then there is the following conclusion

Corollary : Suppose smooth curves $l_1 : x = \alpha(t), y = \beta(t); l_2 : g(x,y) = 0$ are disjoint , when the length

of connecting line $\sqrt{[x-\alpha(t)]^2 + [y-\beta(t)]^2}$ is the shortest ,then

$$\begin{cases} [x-\alpha(t)]\alpha'(t) + [y-\beta(t)]\beta'(t) = 0 \\ [x-\alpha(t)]g_y - [y-\beta(t)]g_x = 0 \end{cases}.$$

The corollary shows that when the connecting line is shortest ,the vector of connecting line is perpendicular to the two direction vectors at conneting points respectively .Thereby ,by solving the system of equation of three variables

$$\begin{cases} [x-\alpha(t)]\alpha'(t) + [y-\beta(t)]\beta'(t) = 0 \\ [x-\alpha(t)]g_y - [y-\beta(t)]g_x = 0 \\ g(x,y) = 0 \end{cases}$$

the coordinates of the two connecting points are obtained .

Especially ,if a curve is straight line (suppose $l_1 : Ax + By = C$) ,by solving the following system of

equations of two variables

$$\begin{cases} \dfrac{g_x}{A} = \dfrac{g_y}{B} \\ g(x,y) = 0 \end{cases},$$

the coordinates of connecting point $Q_2(x_0, y_0) \in l_2 : g(x,y) = 0$ is obtained ,then by solving the following

system of equations of two variables

$$\begin{cases} \dfrac{x-x_0}{A} = \dfrac{y-y_0}{B} \\ Ax + By = C \end{cases}$$

the coordinates of connecting point $Q_1(x,y)$ on the strainght line is obtained.Thereby the distance between

straight line and curve is obtained .

Example 28: Calculate the distance between $l_1 : xy = 32(x, y > 0)$ and $l_2 : 2x + y = 4$.

Solution :Suppose $P(x_1, y_1) \in l_1, Q(x_2, y_2) \in l_2$ .When $|PQ|$ is the shortest ,the normal

vector $N_1 = (y_1, x_1)$ at $P(x_1, y_1)$ is collinear to the normal vector $N_2 = (2,1)$ to straight line l_2 .By

solving the following system of equations

$$\begin{cases} \dfrac{y_1}{2} = \dfrac{x_1}{1} \\ x_1 y_1 = 32 \end{cases},$$

The coordinates of $P(x_1, y_1) = (4,8)$ is obtained .

And then ,the equation of line $PQ : \dfrac{x-4}{2} = \dfrac{y-8}{1}$ is obtained ,by solving the following system of equations

$$\begin{cases} \dfrac{x-4}{2} = y-8 \\ 2x + y = 4 \end{cases}$$

The point $Q(x_2, y_2) = (\dfrac{4}{5}, \dfrac{12}{5})$ on the straight line l_2 is obtained .

Thereby the distance between l_1, l_2 is

$$d = \sqrt{(x_1 - x_2)^2 + (y_1 - y_2)^2} = \sqrt{(4 - \dfrac{4}{5})^2 + (8 - \dfrac{12}{5})^2} = \dfrac{4\sqrt{65}}{5}.$$

Note:If the general equations of the two curves can both be converted into systems of equations with a parameter respectively, namely

$$l_1 : \begin{cases} x = \alpha(t) \\ y = \beta(t) \end{cases}, l_2 : \begin{cases} x = \gamma(s) \\ y = \varphi(s) \end{cases},$$

Then calculating the distance between two curve is changed into solving extreme value of the following function without constraints following question .

$$u(t, s) = \sqrt{[\alpha(t) - \gamma(s)]^2 + [\beta(t) - \varphi(s)]^2}$$

10.2.Distance among point ,curves ,surfaces in space

10.2.1.Distance from point to surface

Theorem 14: Suppose $\Sigma : f(x, y, z) = 0$ is a smooth surface .Fixed point

$P_0(x_0, y_0, z_0) \notin \Sigma$, $Q(x, y, z) \in \Sigma$ .When the length of line segment PQ is the shortest ,the vector of

connecting line $\overrightarrow{P_0O} = (x_0 - x, y_0 - y, z_0 - z)$ is collinear to the normal vector $N = (f_x, f_y, f_z)$ at point Q

in the surface .

The proof of the theorem is similar to the proof of the theorem 12 (omitted) .Thereby by solving the following system of equations

$$\begin{cases} \dfrac{f_x}{x - x_0} = \dfrac{f_y}{y - y_0} = \dfrac{f_z}{z - z_0} , \\ f(x, y, z) = 0 \end{cases}$$

the coordinates of point Q in the surface is obtained .And then distance from point Q to surface is obtained .

Example 29:Calculate the distance from point $P(1,2,1)$ to the conical surface $x^2 + y^2 = z^2$.

Solution :Suppose $Q(x, y, z) \in \Sigma$ .When $|PQ| = \sqrt{(x-1)^2 + (y-2)^2 + (z-1)^2}$ is the shortest ,the normal vector $N = (x, y, -z)$ at point Q is collinear to the vector of connecting line

$\overrightarrow{PQ} = (x-1, y-2, z-1)$. By establishing the following matrix and operating row elementary transformation

$$\begin{pmatrix} x & y & -z \\ x-1 & y-2 & z-1 \end{pmatrix} \rightarrow \begin{pmatrix} x & y & -z \\ 1 & 2 & 2z+1 \end{pmatrix},$$

The following system of equations

$$\begin{cases} y = 2x \\ x(2z+1) = -z, \\ x^2 + y^2 = z^2 \end{cases}$$

is obtained ,namely ,

$$\begin{cases} y = 2x \\ x(2z+1) = -z \\ z = \sqrt{5}x \end{cases} , \quad \begin{cases} y = 2x \\ x(2z+1) = -z \\ z = -\sqrt{5}x \end{cases}$$

The following three suspicious points are obtained

$$Q_1(0,0,0), Q_2(-\frac{1+\sqrt{5}}{2\sqrt{5}}, -\frac{1+\sqrt{5}}{\sqrt{5}}, -\frac{1+\sqrt{5}}{2}), Q_3(\frac{1-\sqrt{5}}{2\sqrt{5}}, \frac{1-\sqrt{5}}{\sqrt{5}}, \frac{\sqrt{5}-1}{2}).$$

Distance from $P(1,2,1)$ to Q_1, Q_2, Q_3 are respectively

$$|PQ_1| = \sqrt{(0-1)^2 + (0-2)^2 + (0-1)^2} = \sqrt{6},$$

$$|PQ_2| = \sqrt{(-\frac{1+\sqrt{5}}{2\sqrt{5}}-1)^2 + (-\frac{1+\sqrt{5}}{\sqrt{5}}-2)^2 + (-\frac{1+\sqrt{5}}{2}-1)^2} = \sqrt{15+3\sqrt{5}},$$

$$|PQ_3| = \sqrt{(\frac{1-\sqrt{5}}{2\sqrt{5}}-1)^2 + (\frac{1-\sqrt{5}}{\sqrt{5}}-2)^2 + (\frac{\sqrt{5}-1}{2}-1)^2} = \sqrt{\frac{30-9\sqrt{5}}{2}}$$

By comparision, $Q_3(\frac{1-\sqrt{5}}{2\sqrt{5}}, \frac{1-\sqrt{5}}{\sqrt{5}}, \frac{\sqrt{5}-1}{2})$ is minimum point ,so the distance from $P(1,2,1)$ to the conical

surface is $\sqrt{\frac{30-9\sqrt{5}}{2}}$.

10.2.2.Distance from point to curve in space

Theorem 15:Suppose $l: \begin{cases} f(x,y,z) = 0 \\ g(x,y,z) = 0 \end{cases}$ is a smooth curve in space .Fixed point

$P_0(x_0, y_0, z_0) \notin l, Q(x, y, z) \in l$ .When the length of line segment PQ is the shortest ,

$$\begin{vmatrix} f_x & f_y & f_z \\ g_x & g_y & g_z \\ x-x_0 & y-y_0 & z-z_0 \end{vmatrix} = 0 .$$

Proof : The following two functions

$$|P_0Q| = \sqrt{(x_0-x)^2 + (y_0-y)^2 + (z_0-z)^2} \quad,$$

$$u(x,y,z) = (x_0-x)^2 + (y_0-y)^2 + (z_0-z)^2$$

have the same extreme points .Under the equality constraints

$$\begin{cases} f(x,y,z) = 0 \\ g(x,y,z) = 0 \end{cases} ,$$

when function $u(x,y,z) = (x_0-x)^2 + (y_0-y)^2 + (z_0-z)^2$ obtains minimum value , $|P_0Q|$ is the shortest .

The equation $f(x,y,z) = 0$ denotes a surface whose normal vector is $N_1 = (f_x, f_y, f_z)$,the equation $f(x,y,z) = 0$ also denotes a surface whose nomal vector is $N_2 = (g_x, g_y, g_z)$, function $u(x,y,z) = (x_0-x)^2 + (y_0-y)^2 + (z_0-z)^2$ denotes a moving surface whose normal vector is $\overrightarrow{P_0O} = (x_0-x, y_0-y, z_0-z)$.

According to theorem 3 ,when function $u(x,y,z)$ obtains extreme value , the above three vectors are coplanar at the extremem point $Q \in l$,thereby

$$\begin{vmatrix} f_x & f_y & f_z \\ g_x & g_y & g_z \\ x-x_0 & y-y_0 & z-z_0 \end{vmatrix} = 0 .$$

This theorem provides a method of calculating the distance from the point $P_0(x_0, y_0, z_0)$ to curve

$l : \begin{cases} f(x,y,z) = 0 \\ g(x,y,z) = 0 \end{cases}$,namely ,by solving the following system of equations

$$\begin{cases} \begin{vmatrix} f_x & f_y & f_z \\ g_x & g_y & g_z \\ x-x_0 & y-y_0 & z-z_0 \end{vmatrix} = 0 \\ f(x,y,z) = 0 \\ g(x,y,z) = 0 \end{cases}$$

obtain suspicious points ,calculate and compare all suspicious extreme values to obtain distance .

10.2.3.Distance between two surfaces

Theorem **16:**Suppose smooth surfaces $\Sigma_1 : f(x,y,z) = 0, \Sigma_2 : g(x,y,z) = 0$ are

disjoint , $P(x_1,y_1,z_1) \in \Sigma_1, Q(x_2,y_2,z_2) \in \Sigma_2$ .When line segment PQ is the shortest ,the three vectors

$(f_{x_1}, f_{y_1}, f_{z_1}), (x_1 - x_2, y_1 - y_2, z_1 - z_2), (g_{x_2}, g_{y_2}, g_{z_2})$ are collinear .

Proof : The following two function of six variables

$$| PQ | = \sqrt{(x_1 - x_2)^2 + (y_1 - y_2)^2 + (z_1 - z_2)^2} \quad .$$

$$u(x_1, x_2, y_1, y_2, z_1, z_2) = (x_1 - x_2)^2 + (y_1 - y_2)^2 + (z_1 - z_2)^2$$

have same extreme points .Under the equality constraints

$$f(x_1, y_1, z_1) = 0, g(x_2, y_2, z_2) = 0$$

When function $u(x_1, x_2, y_1, y_2, z_1, z_2)$ obtains minimum value , $|PQ|$ is the shortest .According to theorem

6 ,when function $u(x_1, x_2, y_1, y_2, z_1, z_2)$ obtains extreme value ,the following three vectors

$$N = (x_1 - x_2, y_1 - y_2, z_1 - z_2, x_2 - x_1, y_2 - y_1, z_2 - z_1) ,$$

$$N_1 = (f_{x_1}, f_{y_1}, f_{z_1}, 0, 0, 0),$$
$$N_2 = (0, 0, 0, g_{x_2}, g_{y_2}, g_{z_2}).$$

are linearly dependent ,so there exist real numbers a, b ,so that $N = aN_1 + bN_2$,namely

$$(x_1 - x_2, y_1 - y_2, z_1 - z_2, x_2 - x_1, y_2 - y_1, z_2 - z_1) =$$

$$a(f_{x_1}, f_{y_1}, f_{z_1}, 0, 0, 0) + b(0, 0, 0, g_{x_2}, g_{y_2}, g_{z_2}) = (af_{x_1}, af_{y_1}, af_{z_1}, bg_{x_2}, bg_{y_2}, bg_{z_2}) ,$$

$$(x_1 - x_2, y_1 - y_2, z_1 - z_2) = a \ (f_{x_1}, f_{y_1}, f_{z_1}) , (x_2 - x_1, y_2 - y_1, z_2 - z_1) = b \ (g_{x_2}, g_{y_2}, g_{z_2}) ,$$

$$(x_1 - x_2, y_1 - y_2, z_1 - z_2) = a \ (f_{x_1}, f_{y_1}, f_{z_1}) = -b(g_{x_2}, g_{y_2}, g_{z_2}) .$$

Thereby ,in order to calculate the distance between the following two smooth surface ,

$$\Sigma_1 : f(x,y,z) = 0, \Sigma_2 : g(x,y,z) = 0$$

We first establish the two following matrixs

$$\begin{pmatrix} f_{x_1} & f_{y_1} & f_{z_1} \\ x_1 - x_2 & y_1 - y_2 & z_1 - z_2 \end{pmatrix}, \begin{pmatrix} x_1 - x_2 & y_1 - y_2 & z_1 - z_2 \\ g_{x_2} & g_{y_2} & g_{z_2} \end{pmatrix},$$

thus we obtain two second-order subdeterminants respectively ,there are four subdeterminants .Let the four
subdeterminants be zero ,we can obtain four equations .And then we solve system of equations to obtain extreme
points ,we finally compare the extreme values to obtain the distance between the two surface .

Especially ,if a surface of the two surfaces is a plane ,when connecting line is the shortest ,normal vector of the other surface is colliear to normal vecor to the plane .Suppose $\Sigma_1 : Ax + By = C$,then by solving the following system of equations

$$\begin{cases} \dfrac{g_x}{A} = \dfrac{g_y}{B} = \dfrac{g_z}{C} , \\ g(x,y,z) = 0 \end{cases}$$

We can obtain the coordinates of point $Q_2(x_0,y_0,z_0) \in \Sigma_2$;and then by solveing the following system of equations

$$\begin{cases} \dfrac{x - x_0}{A} = \dfrac{y - y_0}{B} = \dfrac{z - z_0}{C} \\ Ax + By + Cz = 0 \end{cases}$$

to obtain the coordinates of the point Q_1 in the plane ,we finally obtain the distance .

Example 30 : Calculate the distance between

$$\pi : 2x + y + 3z = 4 \quad \text{and} \quad \Sigma : xyz = 36(x,y,z > 0) .$$

Solution :Suppose $P(x_1,y_1,z_1) \in \pi, Q(x_2,y_2,z_2) \in \Sigma$ .The normal vector of Σ is $N_1 = (2,1,3)$,and the normal vector of Σ is $N_2 = (y_2 z_2, x_2 z_2, x_2 y_2)$ .When connecting line $|PQ|$ is the shortest , the two normal vector are collinear .Namely ,

$$\frac{y_2 z_2}{2} = \frac{x_2 z_2}{1} = \frac{x_2 y_2}{3} ,$$

$$\frac{x_2 y_2 z_2}{2x_2} = \frac{x_2 y_2 z_2}{y_2} = \frac{x_2 y_2 z_2}{3z_2} ,$$

$$\frac{36}{2x_2} = \frac{36}{y_2} = \frac{36}{3z_2} ;$$

Let

$$2x_2 = t, y_2 = t, 3z_2 = t ,$$

then plug them into the equation of Σ

$$\frac{t}{2} \cdot t \cdot \frac{t}{3} = 36 , t = 6 ,$$

obtain $Q(x_2,y_2,z_2) = (3,6,2)$.

Normal equation at $Q(3,6,2)$ in Σ ,namely ,the equation of connecting line PQ is

$$\frac{x-3}{2} = \frac{y-6}{1} = \frac{z-2}{3} ,$$

Let $x = 2s + 3, y = s + 6, z = 3s + 2$,and plug them into equation of π ,

Obtain $P(x_1, y_1, z_1) = (1,5,-1)$.

Thereby the distance between is

$$d = \sqrt{(1-3)^2 + (5-6)^2 + (-1-2)^2} = \sqrt{14} \ .$$

10.2.4. Distance between curve and surface in space

Theorem 17: Suppose smooth curve and smooth surface

$$l : \begin{cases} f(x,y,z) = 0 \\ g(x,y,z) = 0 \end{cases}, \Sigma : F(x,y,z) = 0$$

are disjoint ,points $P(x_1, y_1, z_1) \in l, Q(x_2, y_2, z_2) \in \Sigma$. When

$$|PQ| = \sqrt{(x_1 - x_2)^2 + (y_1 - y_2)^2 + (z_1 - z_2)^2}$$

obtain minimum value ,then the following conditions :

(1) $(f_{x_1}, f_{y_1}, f_{z_1}), (x_1 - x_2, y_1 - y_2, z_1 - z_2), (g_{x_1}, g_{y_1}, g_{z_1})$ are coplane ,

(2) $(x_1 - x_2, y_1 - y_2, z_1 - z_2), (F_{x_2}, F_{y_2}, F_{z_2})$ are collinear.

are simultaneously satisfied .

Proof :When the function of six variables

$$u(x_1, x_2, y_1, y_2, z_1, z_2) = (x_1 - x_2)^2 + (y_1 - y_2)^2 + (z_1 - z_2)^2$$

under the three equality constraints

$$f(x_1, y_1, z_1) = 0, g(x_1, y_1, z_1) = 0, F(x_2, y_2, z_2) = 0$$

obtain minimum value , $|PQ|$ is shortest .Acoording to theorem 6 ,the function obtains extreme value ,the following four six-dimensional vectors

$$N = (x_1 - x_2, y_1 - y_2, z_1 - z_2, x_2 - x_1, y_2 - y_1, z_2 - z_1) ,$$

$$N_1 = (f_{x_1}, f_{y_1}, f_{z_1}, 0, 0, 0),$$
$$N_2 = (g_{x_1}, g_{y_1}, g_{z_1}, 0, 0, 0),$$
$$N_3 = (0, 0, 0, F_{x_2}, F_{y_2}, F_{z_2}).$$

are linearly dependent ,hence there exist real numbers a, b, c ,so that

$$N = aN_1 + bN_2 + cN_3 ,$$

Namely

$$\begin{pmatrix} x_1 - x_2 \\ y_1 - y_2 \\ z_1 - z_2 \\ x_2 - x_1 \\ y_2 - y_1 \\ z_2 - z_1 \end{pmatrix} = a \begin{pmatrix} f_{x_1} \\ f_{y_1} \\ f_{z_1} \\ 0 \\ 0 \\ 0 \end{pmatrix} + b \begin{pmatrix} g_{x_1} \\ g_{y_1} \\ g_{z_1} \\ 0 \\ 0 \\ 0 \end{pmatrix} + c \begin{pmatrix} 0 \\ 0 \\ 0 \\ F_{x_2} \\ F_{y_2} \\ F_{z_2} \end{pmatrix},$$

$$\begin{pmatrix} x_1 - x_2 \\ y_1 - y_2 \\ z_1 - z_2 \\ x_2 - x_1 \\ y_2 - y_1 \\ z_2 - z_1 \end{pmatrix} = \begin{pmatrix} af_{x_1} + bg_{x_1} \\ af_{y_1} + bg_{y_1} \\ af_{z_1} + bg_{z_1} \\ cF_{x_2} \\ cF_{y_2} \\ cF_{z_2} \end{pmatrix},$$

$$\begin{pmatrix} x_1 - x_2 \\ y_1 - y_2 \\ z_1 - z_2 \end{pmatrix} = \begin{pmatrix} af_{x_1} + bg_{x_1} \\ af_{y_1} + bg_{y_1} \\ af_{z_1} + bg_{z_1} \end{pmatrix}, \begin{pmatrix} x_2 - x_1 \\ y_2 - y_1 \\ z_2 - z_1 \end{pmatrix} = c \begin{pmatrix} F_{x_2} \\ F_{y_2} \\ F_{z_2} \end{pmatrix}.$$

So

$$(x_1 - x_2, y_1 - y_2, z_1 - z_2) = a(f_{x_1}, f_{y_1}, f_{z_1}) + b(g_{x_1}, g_{y_1}, g_{z_1}),$$

$$(x_1 - x_2, y_1 - y_2, z_1 - z_2) = -c(F_{x_2}, F_{y_2}, F_{z_2}).$$

Thereby ,when we calculate the distance between smooth curve and smooth surface

$$l: \begin{cases} f(x,y,z) = 0 \\ g(x,y,z) = 0 \end{cases}, \Sigma : F(x,y,z) = 0$$

We first suppose $(x_1, y_1, z_1) \in l, (x_2, y_2, z_2) \in \Sigma$,accoding to condition (1) of theorem 17 ,we can obtian the following equation

$$\begin{vmatrix} x_1 - x_2 & y_1 - y_2 & z_1 - z_2 \\ f_{x_1} & f_{y_1} & f_{z_1} \\ g_{x_1} & g_{y_1} & g_{z_1} \end{vmatrix} = 0,$$

And then according to condition (2) of theorem 17 ,from the following matrix which rank is one

$$\begin{pmatrix} x_1 - x_2 & y_1 - y_2 & z_1 - z_2 \\ F_{x_2} & F_{y_2} & F_{z_2} \end{pmatrix},$$

we can obtain two equations ;we solve system of equations of six variables to obtain extremem points ,we finally compare exreme values to obtain the distance .But if the curve is denoted as a system of equations with a parameter ,then in order to extreme points ,we only solve system of equations .

Corollay :Suppose smooth curve $l: x = \alpha(t), y = \beta(t), z = \gamma(t)$ and smooth surface $\Sigma : F(x, y, z) = 0$

are disjoint , $P \in l, Q \in \Sigma$ .When

$$| PQ |= \sqrt{[x - \alpha(t)]^2 + [y - \beta(t)]^2 + [z - \gamma(t)]^2}$$

obtains minimum value ,the vector $\overrightarrow{PQ} = (x - \alpha(t), y - \beta(t), z - \gamma(t))$ satisfies simultaneously the following conditions :

(1)collinear to normal vector (F_x, F_y, F_z) ,

(2)perpendicular to the direction vector $(\alpha'(t), \beta'(t), \gamma'(t))$ at P on l .

Thereby ,in order to calcaulate distance between curve l and surface $\Sigma : F(x, y, z) = 0$,if curve l can be denoted as the following system of equations with a parameter ,suppose

$$l : x = \alpha(t), y = \beta(t), z = \gamma(t) ,$$

We only solve the following system of equations of four variables

$$\begin{cases} F(x, y, z) = 0 \\ \dfrac{x - \alpha(t)}{F_x} = \dfrac{y - \beta(t)}{F_y} = \dfrac{z - \gamma(t)}{F_z} \\ [x - \alpha(t)]\alpha'(t) + [y - \beta(t)]\beta'(t) + [z - \gamma(t)]\gamma'(t) = 0 \end{cases} ,$$

The extreme points can be obtain ,we finally compare the extreme values to obtain the distance .

Especially ,if l is straight line $: x = mt + a, y = nt + b, z = lt + c$,we only solve the following system of equations

$$\begin{cases} F(x, y, z) = 0 \\ \dfrac{x - mt - a}{F_x} = \dfrac{y - nt - b}{F_y} = \dfrac{z - lt - c}{F_z} \\ m[x - mt - a] + n[y - nt - b] + l[z - lt - c] = 0 \end{cases} ,$$

to obtain extreme points ,we finally compare the extreme values obtain distance .

Example 31:Calcaulate distance between $l : \dfrac{x - 2}{2} = y - 4 = z - 5$ and sphere

$\Sigma : x^2 + y^2 + z^2 = \dfrac{101}{24}$.

Solution :Suppose $P \in l, Q(x, y, z) \in \Sigma$ .Direction vector and the system of equations of line are as follows respectively

$$v = (2,1,1) , x = 2t + 2, y = t + 4, z = t + 5 ;$$

the normal vector to sphere is $N = (x, y, z)$.

Accoding to theorem 17 ,when $|PQ|$ is shortest ,the vector connecting line

$$\overrightarrow{PQ} = (x - 2t - 2, y - t - 4, z - t - 5)$$

is collinear to normal vector $N = (x, y, z)$ in the sphere and is perpendicular to direction vector $v = (2,1,1)$ of

the line ,so the system of equations of four variables

$$\begin{cases} x^2 + y^2 + z^2 = 4 \\ \dfrac{x - 2t - 2}{x} = \dfrac{y - t - 4}{y} = \dfrac{z - t - 5}{z} \\ 2(x - 2t - 2) + (y - t - 4) + (z - t - 5) = 0 \end{cases}$$

is obtained ,namely

$$\begin{cases} x^2 + y^2 + z^2 = 4 \\ \dfrac{2t + 2}{x} = \dfrac{t + 4}{y} = \dfrac{t + 5}{z} \\ 2(x - 2t - 2) + (y - t - 4) + (z - t - 5) = 0 \end{cases} \quad,$$

Let

$$\frac{2t + 2}{x} = \frac{t + 4}{y} = \frac{t + 5}{z} = \frac{1}{m} \quad,$$

Then

$$x = m(2t + 2), y = m(t + 4), z = m(t + 5) \quad,$$

Eliminate unknowns x, y, z ,obtain

$$2[m(2t + 2) - 2t - 2] + m(t + 4) - t - 4 + m(t + 5) - t - 5 = 0,$$

Namely

$$(m - 1)(6t + 13) = 0.$$

If $m = 1$,then $x = 2t + 2, y = t + 4, z = t + 5$,

Plug them into equation of sphere ,obtain the following quadratic equation

$$(2t + 2)^2 + (t + 4)^2 + (t + 5)^2 = \frac{101}{24},$$

But ,this quadratic equation has no real number solution ,so $m \neq 1, 6t + 13 = 0$.

Plug $t = -\dfrac{13}{6}$ into system of equations of line ,obtain $P(-\dfrac{7}{3}, \dfrac{11}{6}, \dfrac{17}{6}) \in l$.

Plug $(x, y, z) = (-\dfrac{7}{3}m, \dfrac{11}{6}m, \dfrac{17}{6}m)$ into equation of sphere ,obtain

$$m^2(\frac{49}{9} + \frac{121}{36} + \frac{289}{36}) = \frac{101}{24},$$

$$m = \pm\frac{1}{2},$$

Then obtain $Q_1(x, y, z) = (-\dfrac{7}{6}, \dfrac{11}{12}, \dfrac{17}{12}) \in \Sigma, Q_2 = (\dfrac{7}{6}, -\dfrac{11}{12}, -\dfrac{17}{12}) \in \Sigma$

Calculate and compare

$$|PQ_1| = \sqrt{(-\frac{7}{6}+\frac{7}{3})^2 + (\frac{11}{12}-\frac{11}{6})^2 + (\frac{17}{12}-\frac{17}{6})^2} = \frac{\sqrt{606}}{12},$$

$$|PQ_2| = \sqrt{(\frac{7}{6}+\frac{7}{3})^2 + (-\frac{11}{12}-\frac{11}{6})^2 + (-\frac{17}{12}-\frac{17}{6})^2} > \frac{\sqrt{606}}{12}.$$

Thereby the distance is $\frac{\sqrt{606}}{12}$.

Because sphere and straight line are exceptional surface and exceptional curve respectively ,so there exist simple method of calculation .Namely ,find out the sphere among the family of spheres $\{\Sigma : x^2 + y^2 + z^2 = u^2, u > 0\}$,which is tangent to the line .The difference between the radius of the tangent sphere and Σ is the distance .That can be done by solving minimum value of function

$u = x^2 + y^2 + z^2$ under equality constraints $\frac{x-2}{2} = y - 4, y - 4 = z - 5$.

The main purpose is demonstrate how to solve system of nonlinear equations here.

10.2.5.Distance between two curves in the space

Theorem 18 :Suppose two smooth curves

$$l_1 : \begin{cases} f(x,y,z) = 0 \\ F(x,y,z) = 0 \end{cases} , l_2 : \begin{cases} g(x,y,z) = 0 \\ G(x,y,z) = 0 \end{cases}$$

are disjoint ,points $P(x_1, y_1, z_1), Q(x_2, y_2, z_2)$ .When line segment $|PQ|$ is the shortest , the following two conditions

1)vectors $(f_{x_1}, f_{y_1}, f_{z_1}), (x_1 - x_2, y_1 - y_2, z_1 - z_2), (F_{x_1}, F_{y_1}, F_{z_1})$ are coplanar,

2)vectors $(g_{x_2}, g_{y_2}, g_{z_2}), (x_1 - x_2, y_1 - y_2, z_1 - z_2), (G_{x_2}, G_{y_2}, G_{z_2})$ are coplanar.

must be satisfied simultaneously .

The proof of the theorem is similar to the proof of previous theorems (omited) .Thereby ,finding out the coordinates of the line endpoints when connecting line is the shortest can be converted into solving the following

system of six variables $l_1 : \begin{cases} f(x_1, y_1, z_1) = 0 \\ F(x_1, y_1, z_1) = 0 \\ \begin{vmatrix} f_{x_1} & f_{y_1} & f_{z_1} \\ x_1 - x_2 & y_1 - y_2 & z_1 - z_2 \\ F_{x_1} & F_{y_1} & F_{z_1} \end{vmatrix} = 0 \\ g(x_2, y_2, z_2) = 0 \\ G(x_2, y_2, z_2) = 0 \\ \begin{vmatrix} g_{x_2} & g_{y_2} & g_{z_2} \\ x_1 - x_2 & y_1 - y_2 & z_1 - z_2 \\ G_{x_2} & G_{y_2} & G_{z_2} \end{vmatrix} = 0 \end{cases}$

.

Note:But if the two curves can both be a system of equations with a parameter respectively ,suppose

$l_1 : x = x(t), y = y(t), z = z(t)$, $l_2 : x = x(s), y = y(s), z = z(s)$,

Then calculating distance between the two curves is changed into solving the minimum value of the following function of two variables

$u(t,s) = [x(t) - x(s)]^2 + [y(t) - y(s)]^2 + [z(t) - z(s)]^2$.

10.3.Supplementary for calculation of distance

(1)Planar smooth curve and smooth surface :

Suppose $l : f(x,y) = 0$ is a curve .If there exist both $f_x(x,y), f_y(x,y)$ at any point on the curve l, and

one of $f_x(x,y), f_y(x,y)$ at least is not zero at any point ,then l is a smoothe curve .In other words ,if at

some point (x_0, y_0) ,one of $f_x(x_0, y_0), f_y(x_0, y_0)$ does not exist ,or $f_x(x_0, y_0) = f_y(x_0, y_0) = 0$,then the

curve is unsmooth at the point (x_0, y_0) .For instance ,curve $l : (x-2)^2 = y^3$ is unsmooth at point $P(2,0)$.

Analogously, if the surface $\Sigma : f(x,y,z) = 0$ is smooth ,then at any point in the surface ,there always exist

$f_x(x,y,z), f_y(x,y,z), f_z(x,y,z)$,and at least one of them is not zero ;for instance ,the conical surface

$(x-2)^2 + (y+1)^2 = z^2$ is not smooth at point $(2,-1,0)$.

There dose not exist any implicit function in neighbourhood of unsmooth point .

(2)While calculating disrance from a point P_0 to a curve l (or surface Σ) ,if there is unsmooth point $Q \in l$

(or $Q \in \Sigma$) ,then the distance $|P_0Q|$ must be considered ;while calculating distance between two curves l_1, l_2

(or surfaces Σ_1, Σ_2) ,if there is a unsmooth point $Q \in l_1$ (or $Q \in \Sigma_1$),then the distance from Q to l_2 (or Σ_2)
must be also be considered .In addition ,while calculating distance between two curves ,if there is an endpoint on a curve ,then the distance from the endpoint to other curve is considered ;while calculating two distance between two surfaces if there is a border line in a surface ,then the distance between the border line to other surface is also be considered .

(3)There dose not exist distance between curve and its asymptotic line .Such as ,there does not exist distance

berween curve $l_1 : b^2x^2 - a^2y^2 = a^2b^2$ and line $l_2 : bx - ay = 0$.

There does also not exist distance between surface and its asymptotic surface . Such as ,there does not exist

distance between surface $\dfrac{x^2}{a^2} + \dfrac{y^2}{b^2} - \dfrac{z^2}{c^2} = 1$ and surface $\dfrac{x^2}{a^2} + \dfrac{y^2}{b^2} - \dfrac{z^2}{c^2} = 0$.

(4)While calculating distance between curves ,if the system of equations has no solution ,then there exist three cases :two curves are intersectant ,a curve is the asymptotic line of other curve ,a curve is parallel to the asymptotic line of other curve .While calculating distance between surfaces ,if the system of equations has no solution ,then there exist three cases :two surfaces are intersectant ,a surface is the asymptotic surface of other surface ,a surface is parallel to the asymptotic surface of other surface .

(5)So far ,there are no better approaches to solving nonlinear system of equations of n variables ($n \geqslant 4$) .Even though nonlinear system of equations of three variables ,it is very difficult to solve them sometimes .But these work can provide us with an idea of analysis method to calculate distances among curves ,surfaces .

References

[1]XIONG Ming.(2015)Corrct a Wide Spread Conclusion of Cator Set Theory[J]Adances in Pure Mathematics[J].doi:10.4236/amp.2015.59050

[2]LIU Shiquan ,LIANG Lifu.The Application of Larangian Multiplier Method in Generalized Variational Principles and Finite Element Method[J].Journal of Harbin Engineering University.2001,21(1):29- 32.

[3]XIONG Ming .Moving Curve(surface) and Degenerate Transformation for Multi-integral[J].Studies in College Mathematics.doi:10.3969/j.issn.1008-1399.2013.04.0022.

[4]XIONG Ming.Generalization of Outer Product and Application[J].Studis in College Mathematics.doi:10.3969/j.issn.1008-1399.2014.04.018.

[5]Chen Jixiou,YU Chonghua,JIN Lu:Mathematical Analysis[M](volum two).Beijin:Higher Education Press:2003:207.

[6]SUEN Jiayong.Solving Exactly Etremum Under Constraints[J]Studies in College Mathematics,doi.10.3969/j.ssn.1008-1399.2006.02.020

[7]WU Chenyu.A Proof to Larange Multiplier[J].Jounal of Shanxi Datong University(Natural Scien Edition).doi.10.3969/j.ssn.1674-0874.2008.03.004

[8]XIE Minhuei,YUN Ziqiou,YI Fahuai[M].Teaching Materials of Mathematical Analysis (volum two).Beijin:Higher Education Press,2004:224-233.

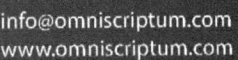